Seeing Animals after Derrida

Ecocritical Theory and Practice

Series Editor: Douglas A. Vakoch, METI

Ecocritical Theory and Practice highlights innovative scholarship at the interface of literary/cultural studies and the environment, seeking to foster an ongoing dialogue between academics and environmental activists.

Recent Titles

Seeing Animals after Derrida edited by Sarah Bezan and James Tink
Eco Culture: Disaster, Narrative, Discourse edited by Robert Bell and Robert Ficociello
Southeast Asian Ecocriticism: Theories, Practices, Prospects edited by John Charles Ryan
Victorian Ecocriticism: The Politics of Place and Early Environmental Justice edited by Dewey W. Hall
Bhopal's Ecological Gothic: Disaster, Precarity and the Biopolitical Uncanny by Pramod K. Nayar
Fallen Animals: Art, Religion, Literature edited by Zohar Hadromi-Allouche
Ecological Thought in German Literature and Culture edited by Gabriele Dürbeck, Urte Stobbe, Hubert Zapf, and Evi Zemanek
The Horse in Literature and Film: Uncovering a Transcultural Paradigm by Francisco LaRubia-Prado
Water in Medieval Literature: An Ecocritical Reading by Albrecht Classen
Sustainability and the City: Urban Poetics and Politics edited by Lauren Curtright and Doris Bremm

Seeing Animals after Derrida

Edited by
Sarah Bezan and James Tink

LEXINGTON BOOKS
Lanham • Boulder • New York • London

Published by Lexington Books
An imprint of The Rowman & Littlefield Publishing Group, Inc.
4501 Forbes Boulevard, Suite 200, Lanham, Maryland 20706
www.rowman.com

Unit A, Whitacre Mews, 26-34 Stannary Street, London SE11 4AB

British Library Cataloguing in Publication Information Available

Library of Congress Cataloging-in-Publication Data

ISBN 978-1-4985-4059-9 (cloth : alk. paper)
ISBN 978-1-4985-4060-5 (electronic)

∞™ The paper used in this publication meets the minimum requirements of American National Standard for Information Sciences—Permanence of Paper for Printed Library Materials, ANSI/NISO Z39.48-1992.

Printed in the United States of America

Contents

Acknowledgments

The editors would like to thank above all Dr. David Coughlan and Elizabeth Wijaya for organizing the seminar "Seeing Animals" for the Annual Meeting of the American Comparative Literature Association in Washington University, Seattle, 2015, at which early versions of some of these of chapters were first presented. We are grateful to Douglas Vakoch, Lindsey Porambo, and Nick Johns at Lexington books for their support for this project. We also extend our thanks to Hollande Bezan, a local Canadian artist, who produced the art for the book cover.

Chapter 12 contains material from *In the Approaches* by Nicola Barker. Published by Fourth Estate, 2014. Copyright © Nicola Barker. Reproduced by permission of the author c/o Rogers, Coleridge & White Ltd., 20 Powis Mews, London W11 1JN.

Introduction

Seeing Animals after Derrida

Sarah Bezan and James Tink

How might we best look at animals following Jacques Derrida? The central theme of Jacques Derrida's seminar *The Animal That Therefore I Am* is the question of how to properly respond to the animal. With the opening anecdote of the naked philosopher facing a cat, and the reflection on the unaccountable mixture of shame and uncertainty which the gaze of this animal provokes in him, Derrida explores in this work, originally delivered in 1997, a proliferation of ways in which philosophy, psychoanalysis, and even the project of Western modernity as such, have chosen to respond or dismiss this relation with the living nonhuman through the category of the animal. The first English publication of the initial work in 2002 (and of the complete seminar in 2008) coincided with other ongoing discussions within critical theory and literary studies around categories of ecocriticism and animal studies, including Derrida's own development of his arguments in the two seminars on the *Beast and the Sovereign*, and his work is now seen as an important intervention of deconstruction (to use that term for Derrida's legacy) in these intellectual fields.[1]

The problem of response or of following is also helpful for thinking about any project, such as this volume, that takes up the matter of Derrida and the visualization of the animal at this point in time. The institution of literary studies and theory can itself seem an act of response, or a critical "turn" to or from an existing network of texts and traces, in which any new critical volume attempts to reorient or review a sense of movement across an intellectual terrain. What then might it mean to reconsider Derrida's work on seeing animals and the limits of the human today, when the "animal turn" in literary studies is already familiar, and furthermore in a period of "deconstruction without Derrida," in which his intellectual legacy faces a new set of questions and objections?[2] Moreover, longstanding schools of ecological and environmental criticism must now be supplemented with not only the controversial idea of the Anthropocene, but also by an entire "non-human" turn in the thought of New Materialism, Speculative Realism, and Object-Oriented Philosophy.[3] Thus, Derrida's work has become part of a wider series of theoretical approaches to the animal and the environment since its inception.[4]

The purpose of this introduction is to therefore try and expose the *dissensus* that surrounds Derrida and the question of seeing animals, visuality, and the exposures of the human; to present the ways in which it has made a critical thinking of the animal possible, without necessarily limiting this to the field of animal studies, environmental criticism, or any category of the posthuman that have signposted subsequent theoretical turns. This is because the relation of the Derrida seminar to the practice of animal studies is in fact a properly problematic and self-questioning one. In order to think about this, we must turn to that original ten-hour lecture/seminar at Cerisy on "The Autobiographical Animal" that later became the published volume and propose, somewhat tendentiously perhaps, a response that follows the animal into a theoretical space.

As is well known, the initial essay "The Animal That Therefore I Am" begins with an account of a banal event: the experience of his being watched while naked and exposed by a domestic cat. This entertaining example of the "abyssal rupture" between the human subject and the animal other in a moment of exposure provides the basis of the extraordinary ten-hour lecture/seminar that follows.[5] Throughout the session, Derrida explores the intellectual uncertainty that the simultaneous proximity and distance of the animal to the human domain has evoked in ethics, philosophy, and psychoanalysis. Yet it should be noted how this entire inquiry is framed by that initial scenario of the animal encounter, in an experience of exposure that is essentially visual, even "face to face," so to speak, as well as naked, vulnerable, and passive. Cary Wolfe identifies this as the core ethical claim that connects Derrida to other philosophers such as Stanley Cavell or Cora Diamond, in that rather than produce a set of rules, they are "confronting our 'exposure' to a permanent condition in which (to use Cavell's phrase) 'there is no way to settle our attitude.'"[6] This exposure also warrants Derrida's own singular investigation into the condition of the animal Other as an event that precedes and enables any category of ethics; what it will not do, to the disappointment of some readers no doubt, is to consider fully what the animal is as such, so much as the implications of the gap or difference the Other-as-animal poses for the philosopher.[7] This is why it is important to recognize Derrida's own variant of animal as the *animot*: a strategic neologism meant to identify that historic "confusion of all nonhuman living creatures within the general and common category of the animal."[8] The term designates the failure or bafflement of culture (Western culture in this case) to account for what it means by animal life, in both terms of the creature or a more general condition of animality. Moreover, Derrida suggests that this failure is in fact the necessary possibility to begin thinking of any form of sentient life.

Derrida's main aim is therefore to challenge philosophy's presumptions of mastery over forms of life that are dismissed as animal in contrast with the human:

> Whenever "one" says "The Animal," each time a philosopher, or anyone else, says "The Animal" in the singular and without further ado, claiming thus to designate every living thing that is held not human (man as *rational animal*, man as political animal, speaking animal, *zōon logon echon*, man who says "I" and takes himself to be subject of a statement that he proffers on the subject of the animal, etc.), well, each time the subject of that statement, this one, this I, does that he utters an *asinanity* [bêtise].[9]

The reason that is so foolish (or bêtise) is because it presumes a calculation or preknowledge of the other, and so attempts to foreclose the event of an encounter with the other (as with the cat) that remains truly abyssal, or even a brief experience of the impossible. What this does *not* mean, however, is a guaranteed continuum or joint identity between the human and any other creature, but to mind the gap that constitutes difference.

> To suppose that I or anyone else for that matter, could ignore that rupture, indeed that abyss, would mean first of all blinding oneself to such contrary evidence: and as far as my own modest case is concerned, it would mean forgetting all the signs that I have managed to give, tirelessly, of my attention to difference, to differences. To heterogeneities and abyssal ruptures as against the homogeneous and the continuous. I have thus never believed in some homogeneous continuity between what calls *itself* man and what *he* calls the animal.[10]

Derrida's intervention can therefore be best understood as questioning overall the historical mediation of being or life into the human and animal, from which all the contributors to this volume proceed. In addition, not only does it have consequences for any eventful encounter with the animal creature, but the possibility of knowledge and ideas of rights. For example, Derrida is reluctant, in a contemporaneous interview with Elisabeth Roudinesco, to identify with Peter Singer's famous project of an animal rights theory (despite his shared admiration for Jeremy Bentham's original question about animal sentience—"Can they suffer?"), as it presupposes a simple extension of anthropocentric jurisprudence over animals irrespective of difference.[11] It is fair to say then that Derrida's argument (as with his comments on the sacrificial basis of carnivorous diets) is by no means an uncomplicated advocacy of popular animal rights theories any more than it is a prognosis of some pure animality within the human.[12] It is instead a thought of the event itself whereby the category of the human is itself exposed as untimely and subject to the other, anterior to the possibility of ethics. To read animals after Derrida is therefore unlike the influential work of Donna Haraway, in which the mediation of human and animal is less a point of exposure than a network of "natureculture" between agents and actors.[13] Rather than this thought of a network or continuum of life, the passivity of an exposure to the other, as demonstrated in the seminar, is that which confounds and complicates any such thought of a normative system.

In consequence, skeptical readers might feel that the animal in Derrida lacks sufficient specificity: some contributors to this volume would want to dispute the status of the animal. Perhaps by way of rebuttal, one should think of the animal as the ongoing problem of the trace of the Other in Derrida's work. Although the topic of animals can in fact be discovered throughout Derrida's published oeuvre, it has been especially associated with the alleged "ethical turn" of a "late Derrida" to pose ongoing questions of *différance* and the trace through ideas of hospitality and justice. The animal needs to be understood alongside concepts of the future such as *l'arrivant,* (that which arrives as an event) or the specter that Derrida had explored in Hamlet and Marx.[14] During the Autobiographical Animal lecture, the example of the cat is itself supplemented by the monstrous Chimera (the animal as intellectual assemblage of living parts that confounds prior knowledge) or as the "divanimality" of the nonhuman, sacred Other.[15] Thus, it partly draws us away from familiar ideas of animality towards the more mysterious, nonhuman precondition of the other and an event, or the nonhuman relation of life to itself. In rejecting both a partition and continuity of human and animal, his work turns to the originating problem of deciding what constituted life, whether that is as phenomenon, ethical category, or material substance.

> Beyond the edge of the so-called human, beyond it but by no means on a single opposing side, rather than "The Animal" or "Animal Life" there is already a heterogeneous multiplicity of the living, or more precisely (since to say "the living" is already to say too much or not enough), a multiplicity of organizations of relations between living and dead, relations of organization or lack of organization among realms that are more and more difficult to dissociate by means of the figures of the organic and inorganic, of life and/or death. These relations are at once intertwined and abyssal, and they can never be totally objectified.[16]

In this case the condition of the living is far from being a state of the human being that could be simply attributed as animality, but instead an idea of an arche-writing, as in an organization of traces and signs as a text that are the preconditions of understanding subjectivity or being.[17] This, after all, provides the nominal theme of the "Autobiographical Animal" session in which it was suggested that any apparent ipseity or sense of self that emerges from writing could be understood as an effect of trace through texts that remain iterable, as if we were following the tracks of some kind of literary animal.[18] Of course, this understanding of the text is essential to Derrida's entire philosophical project, but its restatement in the late Animal session has implications for ideas of the theoretical turn.

This is because we could view the decade since Derrida's death as marked by a concentration of ideas on life, the bioethical and the affect, and indeed ecocriticism, which are sometimes leveled against deconstruction. Agamben and Esposito's studies of bare life and biopolitics may come to mind, and the Deleuzian project of vitalism and immanence provides another perspective for animal studies.[19] As Claire Colebrook has noted, theories based on ideas of life and feeling have emerged as challenges to the alleged insensitivity of the linguistic turn: "Discontent, ill feeling, sentiment, sensitivity, emotion, and affect: all these terms are now championed in ways in which the 'linguistic' prejudices of an earlier era of high theory might be overcome."[20] Furthermore, both Colebrook and Tom Cohen also worry that the tone of Derrida's later work—its own turn, so to speak, to figures of justice, ethics, hospitality—may have itself facilitated its eventual supersession by similar, yet more immediate, discourses of life, ethics, and decision.[21] Certainly it can attract some strong resistance: critical opponent of deconstruction like Slavoj Žižek, for example, can complain that the entire tenor of late-Derrida studies has become too pious and preoccupied with the other in the "usual gang of democracy-to-come-deconstructionist-postsecular-Levinisian-respect-for-Otherness suspects."[22]

At this present point in time, further studies of Derrida and the animal turn may well return and respond to the problem of nonhuman life and the trace in Derrida's career, or find ways to think again about the ethical turn of his later work. John Ó Maoliearca, following François Laruelle's method of nonphilosophy, has explored in what ways the animal could be creatively understood by using and circumventing Derrida, in this case by challenging the imposition of a philosophical decision of difference by Derrida onto the animal.[23] Timothy Morton's interest in environmentalism as a form of unknowable and uncanny textuality attempts to bridge deconstruction with speculative materialist thought, while Colebrook suggests that deconstructive textuality (including animality) might be extended beyond cultural context to a thinking of the entire planetary organization of text.[24] *The Beast and the Sovereign* seminars themselves include discussions of the drive (*Trieb*), auto-affection and ipseity in relation to the question of the survival of life (*survivance*), which remain timely for engaging with affect theory, vitalism, and notions of the world, if not also animal studies.[25] But has this particular line of inquiry, trying to respond to Derrida by following some traces in the essays, led too far away from the ethical priorities of animal studies? What might another approach to *The Animal That Therefore I Am* be like?

To reroute and return ourselves to the present concerns of animal studies, which have installed a series of new critical engagements with the prospect of "seeing animals" in film, visual media, literature, and culture, we might therefore ask *how the human comes to be exposed and made vulnerable in relation to the (in)visible animal.* This ethical exposure of the human through the visual apparatus runs in parallel to what Cary Wolfe would call the dismantling of "the schema of the knowing subject."[26] In this sense, what the human *cannot* possibly see is equally important to what the human *can* see. As we have learned from recent developments in Speculative Realism (Quentin Meillassoux, Timothy Morton, Graham Harman), New Materialism and Material Ecocriticism (Rosi Braidotti, Jane Bennett, Serpil Opperman, and Serenella Iovino), and Neo-Vitalist scholarship (Mel Chen, Eugene Thacker), there is a surplus of knowledge in the more-than-human world of organic and inorganic material life that exceeds the possibilities of our perception. In moving on from metaphor and beyond "the empire of the sign" in postdeconstructionist scholarship,[27] thinkers in animal studies are also denouncing the sacrificial logic that undergirds Derrida's scene of being seen naked by his cat and reconfiguring it in terms of the animal's "real and representational consumption," along with its own agency and in the context of its own material conditions of being.[28] From Nicole Shukin's analysis of the procurement of dead animals in the making of celluloid for film[29] to Giovanni Aloi's[30] and Steve Baker's[31] examinations of dead animals in postmodern art, along with Samuel Alberti's investigation of natural history museums as "mausolea" of postmortem animal specimens,[32] it is clear that animal studies scholarship is continuing to grapple not only with a spectral hauntology in various fields of representation, but also with a visceral and spectacular materiality of living and dead animals. Kelly Oliver's reading of Derrida's "autopsic model" of sovereignty in his essay, "Elephant Eulogy," for instance, reveals how the revenant of a dead elephant "does not appear" and "does not make itself visible and therefore capable of capture or mastery by our senses": it cannot, Oliver maintains, "be brought into the light or placed under the autopsic gaze" in Derrida's paradigm.[33]

The ethical project that therefore emerges in the act of acknowledging the inherent meaning of nonhuman materiality and of the limitations of human perception results in a dramatic shift in scale and perspective. Recognizing the ways in which the human becomes an object in the animal's field of vision,[34] and validating the breadth of nonhuman perception in a diverse range of ecological communities (in the vein of notable German biologist Jakob von Uexküll, for example), we can begin to foster new ways of interpreting the continuous relational processes that connect both human and nonhuman animals. What would it mean, for instance, to go beyond the mammalian eye (i.e., the gaze of a cat) and delve into the realm of the microbe, the reptile, or the insect? If the scale of human perception as we know it is bracketed by the medical and microscopic gaze of post-Enlightenment science on the one hand and of what film studies scholar Anat Pick calls the "ocular inflation" that is the product of "blue planet" images of Earth taken from space on the other,[35] then we are obliged to think through the multiplicitous ways in which our perception of the world distorts our view.

The question, as Kari Weil brilliantly formulates in her book *Thinking Animals* is then: "what sort of visibility [are we] trying to create?" and is it "possible to render nonhuman animals visible without fixing their meanings?"[36] If Giovanni Alois is correct in his determination of the split in animal studies between "two large groups, one of which has embraced posthumanism and its distrust for visibility as a truth-constructing tool and one which is inherently reliant on visibility as a main epistemological[37] then this volume attempts to negotiate the tensions of both camps of thinking in light of the material production of art, its visibilities and invisibilities, and the consideration of the reciprocity (or lack thereof) of the gaze. In doing so, we hope to further challenge the ocularcentrism that is inherent to the conception of the animal-human relationship. Our collection traces the move in posthumanist theory that would build new proximities with the animal in, and through, and at times in spite of, the visual apparatus. The essays in this volume, through their attention to the figuration of animal experience, the exploration of alternative senses (the olfactory, dance) and perspectives (underwater, in space), an exchange of technological gazes in art history and theory, and via new theorizations of time, history, and being, each impart new alternatives to circumventing and reconfiguring Derrida's enduring and poignant reflections on animal life.

The first part of this volume contains some specific responses to Derrida's theoretical project. In his chapter, Gavin Rae attends to the particularities of Derrida's extensive engagement with law and animality, looking in particular to the symbolism of the wolf and Derrida's deconstruction of the law of *versus*. In combating binarism, Rae argues, Derrida is able to probe into the otherness of sovereignty, bringing clarity to the ways in which sovereign authority grounds itself in and through its operation. This operative sovereignty in turn "calls into question straightforward physical-symbolic, real-imaginary divisions," revealing the doubled and interconnected nature of sovereignty and animality. Rae concludes that Derrida's use of wolf imagery is resolutely multiple: "the beast and the human do not form a perfectly symmetrical relationship; they are not two independent, equal monads facing one another. Their relation is moving, dynamic, and asymmetric." David Brooks's,

"The Loaded Cat" offers a renewed and careful analysis of Derrida's encounter with his cat as a philosophical problem: an Oedipal riddle punctuated by castration anxiety, human-animal hybridity (the Sphinx), and the imprecisions of language. The cat, according to Brooks's interpretation of Derrida's lecture, is "loaded" with linguistic signifiers and literary allusions, making it difficult for us to truly see the cat not as a figure or metaphor but as a real animal being, "loaded with herself, her suffering, the weight and intensity of her own existence."

In the second section, contributors explore notions of the trace of the animal in relation to death and forms of memory as forms of "Posthumous Encounters." José Alaniz writes in his chapter, "'The Most Famous Dog in History': Mourning the Animot in Abadzis's Laika" that Laika, a Moscow mongrel stray, is an exemplary case study for thinking through grief, commemoration, and cross-species affectual encounters. Closely attending to the political and aesthetic considerations of the Soviet Union and the "afterlives" of Laika in her various posthumous representations in comics, Alaniz expounds on Derrida's observation that the animal occupies a space of hyper-(in)visibility due to its perceived radical otherness. Abadzis's *Laika* perpetuates this hyper-(in)visibility through its graphic representation of Laika's suffering (which is observed through the eyes of human readers), but Alaniz makes a compelling argument for rereading Derrida's insight into language and visibility. Sarah Bezan's "The Anterior Animal: Derrida, Deep Time, and the Immersive Vision of Paleoartist Julius Csotonyi" advances a renewed critique of deep time in Derrida's thought. Identifying the fossil as a site of common ground between the linguistic and material turns, this chapter examines how Csotonyi's hyper-realistic paleoart formally subverts the primacy of human vision and enables a critique of anteriority. By utilizing immersive and embedded perspectives, indiscernible foregrounds, shifting frames, and expansive scales, Csotonyi carefully composes an image of the paleontological object that tests—and even embraces—the limits of what is possible to validate through human perception. Nicole Mennell's chapter, "'The Dignity of Mankind': Edward Tyson and the Anatomy of a Pygmie" adopts an historicist approach to the question of animals and knowledge suggested by Derrida's discussion of Louis XIV in *The Beast and The Sovereign*, by discussing the work of seventeenth century anatomist and member of the Royal Society, Edward Tyson. Mennell examines his *Anatomy of a Pygmie* (1699), which was the first work in English to categorize the chimpanzee in contrast with the human, and explores how the text attempted to draw a conceptual borderline between the human and the animal, and what this suggests about the development of scientific discourse in the period.

In the third part, "Beyond Ocularcentrism," contributors explore the question for any project of "seeing animals" by discussing the possibilities of other sensory, haptic, or otherwise affective experiences of contact with animal in cinema and fiction. Bonnie Gill's chapter on the French filmmaker Chris Marker explores the alter egos of camera and cat in Marker's filmography and artistic productions. Gill suggests that the numerous cats of Marker's films "become an alternate eye, a witness to the places and people he sees." As a cinematographical mechanism of self-representation and autobiographical reflection, the cat also facilitates a procedure for witnessing and accounting for political demonstrations in Paris. The band of feline flâneurs in Marker's film, *Chats perchés* (2004), Gill argues, mobilize multiple encounters and contacts. Calling on Derrida and Deleuze, Gill discusses how images and clips of cats allow Marker to bring a new perspective to his films, in which "truth is not just ends but means" (*Le Joli Mai*). David Huebert examines the perils of our preoccupation with the visual sensorium in his chapter, "Scenting Wild: Olfactory Panic and Jack London's Ocular Dogs." Opening up new avenues for considering the proper place of scent in Jack London's fiction, Huebert recovers a vital thread of tracking in Derrida's thinking in order to forge a "queer path of the olfactory" from the earliest studies of sexology to modern animal studies scholarship. In Megan Cannella's "Do Androids Dream of Derrida's Cat?: The Unregulated Emotion of Animals in Philip K. Dick's *Do Androids Dream of Electric Sheep?*," we learn that Dick's iconic novel of technological imbrication presages the collapse of life, leaving animals as the only remaining vestige of pure emotion and empathy. Reading Derrida's insight into the nudity of the animal as an expression of affect in "The Animal That Therefore I Am," Cannella's chapter demonstrates how the visibility of animal life is commodified, manufactured, and distorted in a technological age. It is only through the pure, unmitigated transmission of animal emotion in Dick's novel, Cannella insists, that the animal recovers the potency of its autonomous agency.

In the final section "New Arrivals," chapters discuss implications of Derrida's work for rethinking literary and artistic genres. In "Be/Holding each Other: Transgeneic Invisibilities, Anomaly, and Subjectivity in the GFP Bunny Project," Malin Palani considers the changing relationship of the human and nonhuman in the era of biotechnology by discussing the transgenic art of Eduardo Kac. Kac's GFP Bunny project, involving the transfusion of DNA from a Jellyfish into a rabbit and then of the rabbit's DNA into the artist himself, is shown to elucidate many issues pertinent to Derrida's concept of following the animal. In particular, Palani argues, the relationship of the visual and the tactile in Kac's work, whereby we may "be/hold" the transgenic animal, allows us to better understand the importance of affect and passivity that are implied in Derrida's work. Kirsten Strom's chapter, "The Surreal Gaze of the Animal Other: Uncanny Encounters in Magritte and Buñuel" reconsiders the topic of visuality and the uncanny experience of the gaze of the animal as evoked in Derrida's "Animal" essay. It addresses the theoretical problem of seeing animals through artistic representation by considering the tradition of surrealist art in both painting and cinema, and proposes a new lineage of the "Darwinian uncanny" in modern art. Rodolfo Piskorski's "Becoming-Animal and the Two Meanings of Animality: A Derridean Reading of *Black Swan*" provides a theoretically attentive analysis of Darren Aronofsky's acclaimed film *Black Swan* (2010). This psychological drama about rival ballerinas rehearsing a production of *Swan Lake* explores ideas of animality at the level of cultural trope and a more singular artistic fantasy of metamorphosis. Piskorski introduces arguments from Deleuze and Guattari to the discussion of the visual, in order to suggest ways in which the film allows us to reconsider the concept of "becoming animal" in relation to film, materiality, and Derrida's concept of the trace. James Tink's chapter "Approaching Apocalypse: The Typology of Animals in Nicola Barker's *In the Approaches*" explores ways of reading contemporary fiction in terms of Derrida's ideas of "divinanimality." Nicola Barker's 2014 novel *In the Approaches* explores ideas of faith, knowledge, and mysticism as part of a broad romantic comedy set in 1980s England, in which animals played some significant roles in the narrative as agents and even narrators. The chapter examines the uses of animals in the novel in relation to ideas of futurity, apocalypse, and the events that are explored by Derrida's work through the figure of a typology of animals. It proposes a reading of the novel as an example of what critic Nicholas Royle calls a "literary animal," which allows us to consider the representation of both the human and animal as examples a "post-secular" condition of futurity.

NOTES

1. Jacques Derrida, *The Animal That Therefore I Am*, ed. Marie-Louise Mallet, and translated by David Wills (New York, NY: Fordham University Press, 2008). The opening essay "The Animal That Therefore I Am" was first published in English in *Critical Inquiry* 28 (Winter 2002): 369–418. Derrida's seminars on the Beast and the Sovereign are translated in two volumes: *The Beast and the Sovereign*, ed. Michael Lisse, Marie-Louise Mallet, and Ginette Michaud, and translated by Geoffrey Bennington: Vol. I (Chicago, IL: Chicago University Press, 2009); Vol. II (Chicago, IL: Chicago University Press, 2011). Important commentaries on all these seminars include Michel Naas, *The End of the World and Other Teachable Moments: Jacques Derrida's Final Seminar* (New York, NY: Fordham, 2015), and David Farrell Krell, *Derrida and Our Animal Others: Derrida's Final Seminar, "The Beast and the Sovereign"* (Bloomington, IN: Indiana University Press, 2013).

2. Martin McQuillan, *Deconstruction without Derrida* (London: Bloomsbury, 2012) 1–12.

3. Timothy Clark, "What on the World is the Earth? The Anthropocene and Fictions of the World," *The Oxford Literary Review* 35, no. 1 (2013) 5–24.

4. For some overview, see Richard Grusin, ed., *The Nonhuman Turn* (Minneapolis, MN: University of Minnesota Press, 2015), vii–xxix.

5. Derrida, *Animal,* 31.

6. Cary Wolfe, "Exposures," in Stanley Cavell, Cora Diamond, John McDowell, Ian Hacking, and Cary Wolfe, *Philosophy and Animal Life* (New York, NY: Columbia University Press, 2009), 16.

7. Donna Haraway, "When Animals Look Back," *Antennae*, 13 (2010), 5–6.

8. Derrida, *Animal*, 48.

9. Ibid., 31.

10. Ibid., 30.

11. Jacques Derrida and Elisabeth Roudinesco, *For What Tomorrow ... A Dialogue*, translated by Jeff Fort (Stanford, CA: Stanford University Press, 2004), 63. Derrida's discussion of Bentham is in *Animal*, 27.

12. See Martin McQuillan, "Does Deconstruction imply Vegetarianism?" in *Derrida Now: Current Perspectives in Derrida Studies*, ed. John W.R. Phillips (Cambridge, MA: Polity, 2016), 132–159.

13. Donna Haraway, "The Companion Species Manifesto," in *Manifestly Haraway* (Minneapolis, MN: University of Minnesota Press, 2016), 154.

14. Derrida, *Specters of Marx: The State of Debt, the Work of Mourning and the New International*, translated by Peggy Kamuf, (New York, NY: Routledge, 2006), 210–211. On the matter of an ethical turn in late Derrida, see Nicole Anderson, "Ethics: An (ir)Responsibility," in *Jacques Derrida: Key Concepts*, ed. Claire Colebrook (London: Routledge, 2015), 52–53. For an earlier inquiry into animality by Derrida, see *Of Spirit: Heidegger and the Question*, translated by Geoffrey Bennington and Rachel Bowlby (Chicago, IL: Chicago University Press, 1989).

15. Derrida, *Animal,* 41–46; 132.

16. Ibid., 31.

17. "Arche-writing," or the precondition of writing and language is central to the project of "early" Derrida: see *Of Grammatology*, translated by Gayatri Chakravorty Spivak (Baltimore, MD: John Hopkins University Press, 1976). On the concept of arche-writing and life more generally, see Martin Hägglund, *Radical Atheism: Derrida and the Time of Life* (Stanford, CA: Stanford University Press, 2008), 51.

18. Derrida, *Animal*, 50.

19. Giorgio Agamben, *The Open: Man and Animal*, trans. Kevin Attell (Stanford: Stanford University Press, 2004); Roberto Esposito, *Living Thought: The Origins and Actuality of Italian Philosophy*, translated by Zakiya Hanafi (Stanford, CA: Stanford University Press, 2012).

20. Claire Colebrook, "The calculus of individual worth," in *Theory and the Disappearing Future: On de Man, On Benjamin*, ed. Tom Cohen, Claire Colebrook and J. Hillis Miller (London: Routledge, 2012), 143.

21. Tom Cohen, "Toxic Assets: de Man's remains and the ecocatastrophic imaginary (an American fable)" in *Theory and the Disappearing Future*, 106. Claire Colebrook, "Not Symbiosis, Not Now: Why Anthropogenic Change Is Not Really Human," in *The Oxford Literary Review*. 34. no. 2 (2012), 194.

22. Slavoj Žižek, *The Parallax View* (Cambridge, MA: MIT Press, 2006), 11.

23. John Ó Maoilearca, *All Thoughts Are Equal: Laruelle and Nonhuman Philosophy* (Minneapolis, MN: University of Minnesota Press, 2015), 201.

24. Timothy Morton, *Ecology Without Nature: Rethinking Environmental Aesthetics* (Cambridge, MA: Harvard University Press, 2007); "Ecology," in *Jacques Derrida: Key Concepts*, ed. Claire Colebrook (London: Routledge, 2015) 45–46. Claire Colebrook, "Not Symbiosis," 202.

25. Derrida, *Beast & Sovereign* I, 180–1; Vol. II, 258–290.

26. Cary Wolfe, "Human, All Too Human," in *PMLA* 124, no.2 (2009): 564–575.

27. Rosi Braidotti, "De-Oedipalizing the Animal Other," *PMLA* 124, no.2 (2009): 526–532.

28. Susan McHugh, "Literary Animal Agents," *PMLA* 124, no.2 (2009): 487–495.

29. Nicole Shukin, *Animal Capital: Rendering Life in Biopolitical Times* (Minneapolis, MN: University of Minnesota Press, 2009).

30. Giovanni Aloi, *Art and Animals* (London: I.B. Tauris, 2012).

31. Steve Baker, *The Postmodern Animal* (London: Reaktion Books, 2000).

32. Samuel Alberti, ed. *The Afterlives of Animals: A Museum Menagerie* (Charlottesville, VA: University of Virginia Press, 2011).

33. In *The Animal Question in Deconstruction*, ed. Lynn Turner (Edinburgh: Edinburgh University Press, 2013).

34. Ron Broglio, "Thinking with Surfaces," in *Animals and the Human Imagination: A Companion to Animal Studies*, eds. Aaron Gross and Anne Vallely (New York, NY: Columbia University Press, 2012), 238–58.

35. Anat Pick, "Three Worlds: Dwelling and Worldhood on Screen," in *Screening Nature: Cinema Beyond the Human*, eds. Anat Pick and Guinevere Narraway (New York, NY: Berghahn, 2013), 21–36.

36. Kari Weil, *Thinking Animals: Why Animal Studies Now?* (New York, NY: Columbia University Press, 2012).

37. Aloi, Giovanni. "Animal Studies and Art: Elephants in the Room" (special editorial of Antennae: Journal of Nature in Visual Culture). pp 1-31 (March 2015): 5-6. Accessed online: http://www.antennae.org.uk/back-issues-2015/4589877799.

WORKS CITED

Agamben, Giorgio. *The Open: Man and Animal*. Translated by Kevin Astell. Stanford: Stanford University Press, 2004.

Alberti, Samuel, ed. *The Afterlives of Animals: A Museum Menagerie*. Charlottesville, VA: University of Virginia Press, 2011.

Aloi, Giovanni. *Art and Animals*. London: I.B. Taurus, 2012.

Anderson, Nicole. "Ethics: An (ir)Responsibility." In *Jacques Derrida: Key Concepts*, edited by Claire Colebrook, 48–57. London: Routledge, 2015.

Baker, Steve. *The Postmodern Animal*. London: Reaktion Books, 2000.

Braidotti, Rosi. "De-Oedipalizing the Animal Other." *PMLA* 124, no.2 (2009): 526–532.

Broglio, Ron. "Thinking With Surfaces." In *Animals and the Human Imagination: A Companion to Animal Studies*, edited by Aaron Gross and Anne Vallely, 238–258. New York, NY: Columbia University Press, 2012.

Clark, Timothy. "What on the World is the Earth? The Anthropocene and Fictions of the World." *The Oxford Literary Review* 35, no. 1 (2013): 5–24.

Cohen, Tom. "Toxic Assets: de Man's Remains and the Ecocatastrophic Imaginary." In *Theory and the Disappearing Future: On de Man, On Benjamin*, edited by Tom Cohen, Claire Colebrook and J. Hillis Miller, 89–129. London: Routledge, 2012.

Colebrook, Claire. "Not Symbiosis, Not Now: Why Anthropogenic Change is not really Human." *Oxford Literary Review* 34, no. 2 (2012): 185–210.

———. "The Calculus of Individual Worth." In *Theory and the Disappearing Future: On de Man, On Benjamin*, edited by Tom Cohen, Claire Colebrook and J. Hillis Miller, 130–152. London: Routledge, 2012.

Derrida, Jacques. *Of Grammatology*. Translated by Gayatari Chakravorty Spivak. Baltimore, MD: John Hopkins University Press, 1976.

———. *Of Spirit: Heidegger and the Question*. Translated by Geoffrey Bennington and Rachel Bowlby. Chicago, IL: Chicago University Press, 1989.

———. *Specters of Marx: The State of Debt, the Work of Mourning and the New International*. Translated by Peggy Kamuf. New York, NY: Routledge, 2006.

———. *The Animal That Therefore I Am*. Edited by Marie-Louise Mallett and translated by David Wills. New York, NY: Fordham University Press, 2008.

———. *The Beast and the Sovereign*. Volume 1. Edited by Michael Lisse, Marie-Louise Mallet, and Ginette Michaud, and translated by Geoffrey Bennington. Chicago, IL: Chicago University Press, 2009.

———. *The Beast and the Sovereign*. Volume 2. Edited by Michael Lisse, Marie-Louise Mallet, and Ginette Michaud, and translated by Geoffrey Bennington. Chicago, IL: Chicago University Press, 2011.

Derrida, Jacques and Elisabeth Roudinesco. *For What Tomorrow ... A Dialogue*. Translated by Jeff Fort. Stanford, CA: Stanford University Press, 2004.

Esposito, Roberto. *Living Thought: The Origins and Actuality of Italian Philosophy*. Translated by Zakiya Hanafi. Stanford, CA: Stanford University Press, 2012.

Grusin, Richard, ed. *The Nonhuman Turn*. Minneapolis, MN: University of Minnesota Press, 2015.

Hägglund, Martin. *Radical Atheism: Derrida and the Time of Life*. Stanford, CA: Stanford University Press, 2008.

Haraway, Donna. "The Companion Species Manifesto." In *Manifestly Haraway*, 3–90. Minneapolis, MN: University of Minnesota Press, 2016.

———. "When Animals Look Back." *Antennae* 13 (2010): 4–9.

Krell, David Farell. *Derrida and Our Animal Others: Derrida's Final Seminar, "The Beast and the Sovereign."* Bloomington, IN: Indiana University Press, 2013.

McHugh, Susan. "Literary Animal Agents." *PMLA* 124, no. 2 (2009): 487–495.

McQuillan, Martin. *Deconstruction without Derrida*. London: Bloomsbury, 2012.

———. "Does Deconstruction imply Vegetarianism?" In *Derrida Now: Current Perspectives in Derrida Studies*, edited by John W.R. Phillips, 132–159. Cambridge, MA: Polity, 2016.

Morton, Timothy. "Ecology." In *Jacques Derrida: Key Concepts,* edited by Claire Colebrook, 41–47. London: Routledge, 2015.
———. *Ecology Without Nature: Rethinking Environmental Aesthetics.* Cambridge, MA: Harvard University Press, 2007.
Naas, Michael. *The End of the World and Other Teachable Moments: Jacques Derrida's Final Seminar.* New York, NY: Fordham University Press, 2015.
Ó Maoilearca, John. *All Thoughts Are Equal: Laruelle and Nonhuman Philosophy.* Minneapolis, MN: University of Minnesota Press, 2015.
Pick, Anat. "Three Worlds: Dwelling and Worldhood on Screen." In *Screening Nature: Cinema Beyond the Human*, edited by Anat Pick and Guinevere Narraway, 21–36. New York, NY: Berghahn, 2013.
Shukin, Nicole. *Animal Capital: Rendering Life in Biopolitical Times.* Minneapolis, MN: University of Minnesota Press, 2009.
Turner, Lynne, ed. *The Animal Question in Deconstruction.* Edinburgh: Edinburgh University Press, 2013.
Weil, Kari. *Thinking Animals: Why Animal Studies Now?* New York, NY: Columbia University Press, 2012.
Wolfe, Cary. "Exposures." In Stanley Cavell, Cora Diamond, John McDowell, Ian Hacking, and Cary Wolfe, *Philosophy and Animal Life*, 1–42. New York, NY: Columbia University Press, 2008.
———. "Human, All Too Human." *PMLA* 124, no. 2 (2009): 564–575.
Žižek, Slavoj. *The Parallax View.* Cambridge, MA: The MIT Press, 2006.

Part I

NEW ORIENTATIONS IN DERRIDA'S PHILOSOPHY

Chapter 1

The Wolves of the World

Derrida on the Political Symbolism of the Beast and the Sovereign

Gavin Rae

Without wanting to sanction the notion of an "early" and "later" Derrida, one that, by positing a rupture within his *oeuvre* would reaffirm the binary logic he criticizes, Jacques Derrida's later writings are distinguished from his earlier ones because of their explicit and extensive engagement with two themes: the nature of law, including its relationship to sovereignty, and the question of animality, including the human-animal relationship. While Derrida deals with these themes in a variety of places, the former most notably in texts, including, but not limited to, the important essays, "Force of Law: The 'Mystical Foundation of Authority'"[1] and "Before the Law,"[2] and the book *Rogues*, and the latter most extensively in the collection of essays translated as *The Animal That Therefore I am*[3] they converge in his final lectures, published in two volumes as *The Beast and the Sovereign*.[4] The first volume, spanning the period from December 12, 2001, to March 27, 2002, undertakes a questioning of the relationship between the notion of the beast and the sovereign that, in showing the intimate conceptual relationship between these two figures, undercuts the binary oppositions between beast and sovereignty and human and animal that have long dominated Western thinking. Engaging with the relationships between (animal) *bestiality* and (human) *sovereignty* and, by extension, *animal* (bestiality) and *human* (sovereignty), depends on, but also brings forth, the questions of what it means to be "animal" and what it means to be "human." The second volume, spanning the following year's lectures from December 11, 2002, to March 26, 2003, deals with these through an unconventional analysis of Martin Heidegger's 1929–1930 lecture course, translated as *The Fundamental Concepts of Metaphysics*, and Daniel Defoe's *Robinson Crusoe*. Bringing together so-called philosophical and literary texts in this way and on this topic not only challenges any understanding that insists on a fundamental and simple division between these disciplines, but also questions long-held assumptions about any straightforward animal-human divide. As I will argue in this chapter, both are manifestations of Derrida's ongoing critique of binary oppositions, a critique that is integral to *The Beast and the Sovereign*.

Relatively little work has been done on either volume of *The Beast and the Sovereign*. David Farrell-Krell's *Derrida and Our Animal Others*[5] was the first monograph to engage with Derrida's final seminar and, while it provides a brief overview of its themes which allows the reader to quickly identify their flow, the discussion *of both volumes* is condensed into only two chapters. As a consequence, there is no in-depth analysis of their content. Michael Naas' *The End of the World and Other Teachable Moments* fares somewhat better on this front in that he does actually undertake an extended, in-depth analysis of the arguments introduced by Derrida, but he "concentrate[s] primarily on the second volume of *The Beast and the Sovereign*"[6] with the consequence that there is little engagement with the first volume and, in particular, the question of the relationship between the beast and the sovereign. This is also found in the scholars who have written on the first volume. While they note the importance of the beast-sovereign relationship, there is a tendency to takes its content for granted to focus on other aspects of the seminar. Jacques de Ville,[7] for example, analyses Derrida's reading of Hobbes, Chris Danta[8] highlights the crucial role that fables play in the first volume, and Mathew Chrulew[9] looks at Derrida's treatment of modernity and the question of biopolitics. These reveal the depth of Derrida's treatment of the topic, but their focus means that they miss aspects of Derrida's treatment of the beast-sovereign relationship, especially the liminality inherent to it.

The purpose of this chapter is to start to rectify this oversight while also challenging Naas' assertion that "the seminars differ greatly from Derrida's published works"[10] to the extent that their gradual publication will result in "what can only be characterised as a wholly *other corpus* besides the one we already know."[11] By distinguishing, in a hard sense, between works originally written for publication and those originally designed for teaching, Naas depends upon and, indeed, reestablishes the binary logic that Derridean deconstruction calls into question. As a consequence, Naas offers a reading of Derrida that depends upon and affirms the very structures that Derrida aims to overcome. My guiding contention, however, is that the content of the seminars is not wholly other than that of the nonseminar works. Derrida's analysis of the entwined relationship between bestiality and sovereignty in the first volume of *The Beast and the Sovereign* challenges any notion of an absolute human/animal divide and so depends upon, while also contributing to, his long-standing analysis of, what he calls in the early influential essay "Dissemination," the "crisis of *versus*."[12] There are two aspects to this crisis: first, that the logic of binary oppositions ("this *versus* that," "pure *versus* impure") has fallen apart under the logical contradictions it throws up. Second, Derrida's affirmation of this disintegration as part of his attempt to undermine a reliance on or return to binary logic. *The Beast and the Sovereign* contributes to this by examining the relationships between animality and bestiality, the human and the animal, and the human, animal, and the juridical-political to show that bestiality and sovereignty have a similar exclusionary relationship to law. While both stand outside of law, the sovereign does so because it transcends law, while the beast does so because it is excluded from the law. Nevertheless, the beast and the sovereign come together through the imagery of the wolf: in the case of the former because the wolf is, all too often, associated with "pure" bestiality and in the case of the latter because the wolf marauding through the world is taken to be emblematic of the way the sovereign is unconstrained by law. As such, Derrida's analysis of the relationship between bestiality, sovereignty, and law in *The Beast and the Sovereign* not only challenges the human/animal divide and so deepens and extends his critique of the logic of "*versus*," but, by extension, also highlights and challenges us to reexamine the political symbolism traditionally associated with the human/animal relationship.

CROSSING BOUNDARIES

Ever sensitive to the vagaries, peculiarities, and subtleties of language, Derrida starts his analysis of the relationship between the beast and the sovereign—"this odd couple" (BS1: 137/191) as he calls it—by pointing out that these two terms are, in French at least, intimately related to the sexual difference. Whereas the simplicity of English means that there is only one definitive article used for both *the* masculine and *the* feminine, French distinguishes between the two: *la* denoting the feminine and *le* the masculine. This is immediately linked to the relationship between the beast and the sovereign by virtue of each possessing different, gendered definitive articles: "the beast [feminine: *la bête*] and the sovereign [masculine: *le souverain*]. *La, le*" (BS1: 1/20). With this, Derrida points out that his analysis of the beast and the sovereign will not simply be concerned with these two terms, but will extend into and depend on an engagement with the way(s) in which gender influences our political and zoological categories and, by extension, political symbols. Indeed, in the essay "Before the Law," Derrida insists on this importance: "in French, the feminine determines a semantic contagion that we cannot *forget*."[13] The political importance of the beast's (French) feminization is seen from the structural analogy between the subjugation of the beast and the subjugation of women, an analogy that affirms the phallogocentrism of Western thinking. As Derrida explains, "the beast was often the living thing to be subjected, dominated, domesticated, mastered, like, by a not insignificant analogy, the woman, the slave, or the child" (BS1: 66/101). The femininity of the (French) beast associates the feminine with the beast and so turns her into a nonhuman other, one to be subjugated for the benefit of the masculine human. This appears to establish a binary opposition between the masculine sovereignty and the subjugated feminine beast, but no sooner has Derrida noticed this than he calls it into question by noting that, on the symbolic level, the masculine human sovereign and the feminine beast are engaged in a relationship of entwinement. Specifically, the beast and the sovereign, feminine and masculine, come together through the political symbolism of the wolf. For this reason,

> one cannot be interested in the relations of beast and sovereign, and all the questions of the animal and the political, of the politics of the animal, of man and beast in the context of the state, the *polis*, the city, the republic, the social body, the law in general, war and peace, terror and terrorism, national or international terrorism, etc., without recognizing some privileging of the figure of the "wolf." (BS1: 9/28–29)

To emphasize his point, Derrida provides examples, taken from different cultures and societies that indicate the political significance of the wolf. These include the fundamental role that the wolf plays in the myth of the foundation of Rome, or how North American Indians, specifically "among the Ojibwa, in certain variants of the story, the hero Manabozho (who most of the time gets on well with the brother) either remains inconsolable at his death or else kills him himself; and his brother, dead or killed by him, is a Wolf: the Wolf" (BS1: 9/29). Similarly, in Nordic and Germanic cultures, when Wotan or Ordin, who "is a warrior god, a god of warlike fury ... sits on the throne, he is flanked by two wolves, who are like the insignia of his majesty, living coats of arms, the living heraldry of his sovereignty" (BS1: 10/30), while, in the myth of Wotan's/Odin's son, Thor, the figure of the "sinister and voracious ... wolf" (BS1: 10/30) is fundamental to the foundation of justice. Then, of course, there is the philosophical importance of the wolf, manifested in the work of Hobbes and Rousseau (BS1: 9–10/29–30), not to mention its importance to psychoanalysis, as manifested in the famous case of the Wolf Man (BS1: 64/99). Indeed, Derrida notes that "the nickname 'wolf' is [often] given to a head of state as Father of the Nation. Mustapha Kemal who [gave] himself the name Atatürk (Father of the Nation) was called the 'grey wolf' by his partisans, in memory of the mythical ancestor Genghis Khan, the 'blue wolf'" (BS1: 17/39). Interestingly, and while Derrida does not note it, David Farrell-Krell[14] gives the example of the American conservative Fox news station to not only show the continuing political importance of wolf imagery, but also suggest that with the advent of new communication and multimedia outlets, the wolf imagery straddles ever more boundaries, politicizing them and, at the same time, showing and cementing its political importance. There are also the examples of Leicester City Football Club, nicknamed "The Foxes," who, against all predictions and reported betting odds of around 5000–1 at the start of the season, won, and so in a sense became sovereign over, the English Premier Football (Soccer) League in season 2015–2016, and, more ominously, "The Wolf of Wall Street" Jordan Belford,[15] who earned his nickname for his highly successful, but ultimately illegal, predatory actions in the stock market. Regardless of culture, geography, and history, it appears then that wolf imagery has imbued and continues to imbue meaning to various aspects of different societies.

There are also *conceptual* reasons why Derrida focuses on the political symbolism of wolves. Specifically, it better allows him to call into question the conceptual boundaries that have long thought to separate the supposedly political human from the nonpolitical animal. The figure of the wolf is particularly applicable to this undertaking because, in its actual and symbolic form, the wolf crosses boundaries and so poses "thorny frontier questions" (BS1: 4/23):

> without asking permission, real wolves cross humankind's national and institutional frontiers, and his sovereign nation-states; wolves out in nature [*dans la nature*] as we say, real wolves, are the same on this side or on the other side of the Pyrenes or the Alps; but the figures of the wolf belong to cultures, nations, languages, myths, fables, fantasies, histories. (BS1: 4–5/23)

Despite their differences, both real and conceptual wolves transgress boundaries. Linking this transgression to a notion of sovereignty does not, however, simply mean that "the social, the political, and in them the value or exercise of sovereignty are merely disguised manifestations of animal force" (BS1: 14/34). Sovereignty cannot simply be reduced to "bestiality or barbarity or inhuman cruelty" (BS1: 14/34). The beast and the sovereign are defined by *différance* and so cannot be collapsed into "one" another.[16] Nor is it sufficient to invert the relation to point out "that the animal is already political" (BS1: 14/34) and defend this by showing that in

> what are called animal societies [there exists] the appearance of refined, complicated organizations, with hierarchical structures, attributes of authority and power, phenomena of symbolic credit, so many things that are so often attributed to and so naïvely reserved for so-called human *culture*, in opposition to *nature*. (BS1: 15/35)

Derrida's analysis of the beast-sovereign relationship does not entail such a superficial, empirical analysis of the ways in which either aspect appears in the other. It focuses on the *logic* underpinning these analyses to show the fallacy inherent to thinking of the beast and sovereign in oppositional terms, a critique that also questions the relationship between humans and animals, law and power, masculinity and femininity, and the physical and symbolic. In so doing, Derrida's analysis of the beast-sovereign relationship depends upon, forms part of, and extends his long-standing deconstruction of the logic of *versus*. This does not entail anything as simplistic and definitive as simply inverting the aspect privileged nor does it entail the collapsing of one aspect into the other. The difference and, thus, distinctness of each aspect is respected, the relationship between them, including the way each influences the other is emphasized, and, crucially, the boundary that separates, unites, and differentiates one from the other is questioned. Through this, "the fragility and porosity of th[e] limit between nature and culture" (BS1: 15/36) is interrogated, probed, and inquired into.

THE OTHERNESS OF SOVEREIGNTY

To probe the beast-sovereign relationship further, Derrida engages with the question of the "nature" of sovereignty. In response, he claims that sovereignty does not refer to size and so cannot be quantified "within the oppositional grid of the large and the small" (BS1: 259/348). Sovereignty refers to a power rather than an attribute, a verb rather than a noun, the "possession" of which allows one to act in a particular way. To develop this, Derrida moves the discussion away from the concept of "sovereignty" to the one through which this power is operative: the sovereign. The sovereign "is alone insofar as he is unique, indivisible, and exceptional" (BS2: 7/30). Indeed, Derrida claims that "sovereignty cannot be shared, it is indivisible" (BS2: 8/30), which does not mean that there is only one sovereign, only that, whenever sovereignty is claimed, it is always indivisible. As Derrida puts it, "the sovereign is alone (sovereign) or is not" (BS2: 8/30). For this reason, the key aspect of sovereignty is "ipseity" (BS1: 214/287), which entails the capacity to decide for oneself from oneself; or, as Derrida puts it, "the sovereign is one who has his end in himself or is the end of everything" (BS1: 346/458).

There is, then, no external foundation or ground to sovereignty; if there were, that ground would be sovereign. Sovereignty is, for Derrida, groundless, which must not be understood in the sense of lack, as in the groundlessness of sovereignty entails a lack of ground. This would trap sovereignty within the ground/groundless binary opposition and so make it dependent on that logic. Sovereignty would not, as a consequence, be sovereign, but dependent. What Derrida has in mind here is a somewhat paradoxical understanding of sovereignty wherein it has no (external) ground, but yet is not simply defined from an arbitrary abyss. We might call it a *groundless ground*: the former because sovereignty emanates from itself not from another and this "itself" is always defined by *différance* meaning it is always differentiating and deferred; the latter because it is the sovereign point that supports all else. Indeed, in the important essay "Force of Law: The 'Mystical Foundation of Authority,'" Derrida explains that

> the origin of authority, the founding or grounding [*la fondacionou le fondemont*], the positing of the law [*loi*] cannot by definition rest on anything but themselves, they are themselves a violence without ground [*sans fondemont*]. This is not to say that they are in themselves unjust, in the sense of "illegal" of "illegitimate." They are neither legal nor illegal in their founding moment. They exceed the opposition between funded and unfounded, or between any foundationalism or antifoundationalism.[17]

Sovereign authority lies at the "foundation" of law in the sense that it is from sovereign authority that law emanates. But, at the same time, this does not establish a binary opposition between itself and law as this would limit sovereign authority within the restricted economy of the binary opposition. Sovereign authority lies outside of this binary logic by virtue of grounding it. Derrida points out, however, that, by grounding itself in and from itself, we run into the aporia of how and why sovereignty grounds itself and, by extension, why and from where authority and, thus, law, emanates. For Derrida, this points to the "mystical foundation"[18] of sovereignty and, by extension, law. Indeed, he points out that the nonrational "ground" entails

> a certain power to *give*, to *make*, but also to *suspend* the law; it is the exceptional right to place oneself above right, the right to non-right ... which runs the risk of carrying the human sovereign above the human, toward divine omnipotence (which will moreover most often have grounded the principle of sovereignty in its sacred and theological origin). (BS1: 16–17/37–38)

Sovereignty is not only "grounded" in mystical, theological foundations, but, through the exercise of sovereignty, *the sovereign* transcends human status towards divine omnipotence. As such, we find that "like God, the sovereign is above the law and above humanity, above everything" (BS1: 57/91). Indeed, Derrida claims that this transcendence appears so strange and other to humans that the sovereign appears to be "a bit stupid (*bête*)" (BS1: 57/91), a claim that plays on the double significance of the French word for beast and stupidity (*bête*) and allows Derrida to hold that the sovereign is, then, a little "like a beast" (BS1: 57/91). Of course, Derrida is aware that the sovereign is not, strictly speaking, God or a beast, but only "*like* a God, *like* a beast" with the consequence that the question of what this means is "still on the table" (BS1: 57/91). While leaving the question of the God-*like*, animal-*like* nature of sovereignty open, Derrida's point is that when perceiving the sovereign/sovereignty, we tend to be exposed to something that appears to be "other-than-human."

We see this further if we turn to Derrida's comments on the masculinity of the (French) sovereign. By pointing out that the phallus is most typically associated with masculinity and the latter to sovereignty, Derrida establishes an analogous link between the phallus and sovereignty. This continues through his insistence that the defining quality of the phallus is its hardness, which is then linked the "machinelike or mechanical" (BS1: 222/297). By noting that the defining feature (i.e., the phallic hardness) of masculine sovereignty shares similarities with the metallic hardness of the nonhuman machine, Derrida questions and, in questioning, suggests that at the heart of human sovereignty lies the nonhuman (BS1: 222/297). Those critical of Derrida will no doubt point out that there are logical leaps in this "argument" and that the analogy he develops between the phallic and machinic hardness is, to say the least, underdeveloped. Indeed, Derrida seems to acknowledge this when he writes with the conditional "*if* there is something in [the phallus's] stiffness and hardness that is machinelike or mechanical" (BS1: 222/297, italics added), but his overall point seems to be to reinforce his insistence that sovereignty/the sovereign are intimately linked to the nonhuman.

THE MEDIATION OF LAW

Having identified an analogy between sovereignty and bestiality, Derrida explores the beast-sovereign relationship further through the mediation of law. The reason being that there is, on Derrida's telling, a structural similarity between the beast's and the sovereign's relationship to law. Specifically, the beast and the sovereign "seem to have in common their being-outside-the-law" (BS1: 17/38) in that they exist "at a distance from or above the laws" (BS1: 17/38) and lack respect for any conception of "absolute law" (BS1: 17/38). In the case of the beast, its supposed savagery means that it is understood to exist outside the "civilizing" structures of law and, by extension, reason. There is, then, a cleavage between bestiality and law, one that means that the former exists beyond the law. The sovereign, as the one who creates the law, is he who stands outside the law. Only this "external" position allows the sovereign to create the law. As the groundless ground of law, the sovereign is the one who instantiates, defines, and supports the law without being supported by that which he creates. The notion of beyond-the-law is not, however, singular. As Derrida explains, the beyond or outside-of-the-law can take two forms:

> on the one hand, (and this is the figure of the sovereign), [it can] take the form of being-above-the laws, and therefore take the form of Law itself, with a capital L, the condition of the law, were before, above, and therefore outside the law,

> external or even heterogeneous to the law; but being-outside-the-law can also, on the other hand (and this is the figure of what is most often understood by animality or bestiality) … situate the place where the law does not appear, or is not respected, or gets violated. (BS1: 17/38)

Derrida's overall point is that being-above-the-law can take on numerous forms: beyond, away-from, the transcendental condition of law, and/or a place from where the law does not appear but from which it emanates and is violated. It is not the case that there is a straightforward inside/outside relationship to the law. There are multiple conceptions of "outside" as well as multiple "insides," a multiplicity mirrored through the possible relationships between each. With this, Derrida again disrupts any straightforward binary opposition between the sovereign/beast "sitting" beyond-the-law and the law itself.

Nevertheless, while both the beast and the sovereign are distinct from the law, neither is defined by a total absence of law. In the case of the sovereign, he is outside by virtue of being a condition of law *and* inside the law by virtue of being legally defined within the legal system. The beast, whose identity is defined in terms of the absence of law, is always associated with that which it lacks and so is always defined from law. What is important, however, is that law clings to the beast and the sovereign by virtue of their exclusion from it. For this reason, Derrida notes that both share a relationship to criminality that also entails an exclusion to law. The conceptual link between sovereignty and criminality allows sovereign states to be associated with criminality under the nomenclature "rogue states," which are, according to Derrida,

> delinquent states, criminal states, states that behave like brigands, like highway robbers or like vulgar rapscallions who just do as they feel, do not respect international right, stay in the margins of international civility, violate property, frontiers, rules and good international manner, including the laws of war … according to the rhetoric of heads of sovereign states who for their part claim to respect international right. (BS1: 18–19/40)

The notion of rogue states is also linked to bestiality in that the rogue animal is the one "who does not even respect the law of the animal community, of the pack, of the horde, of its kind. By its savage or indocile behavior, it stays or goes away from the society to which it belongs" (BS1: 19/40). Paul Patton notes that this association allows Derrida to move from a purely conceptual analysis of the beast-sovereign to one that enters the practical arena of international relations and, in particular, to discussions of those events that "were current at the time of these lectures, which began in December 2001, shortly after the attack on the World Trade Center on September 11, and continued through the lead-up to the invasion of Iraq in 2003."[19] With this, Derrida problematizes the theoretical-practical division by showing how they interact with, interrogate, and support one another. To mention just one particularly pertinent example, he notes that Saddam Hussein became known as the "Beast of Baghdad" (BS1: 20/42), a nomenclature that affirms the intimate connection between bestiality and sovereignty and calls into question straightforward physical-symbolic, real-imaginary divisions: the real, physical sovereign is seen and understood through reference to the imaginary, symbolic constructions associated with bestiality, which are influenced by exposure to actual, physical sovereigns and beasts. The result is, on Derrida's telling, that the sovereign's actual status and position depend on and are reinforced by the perceived defining features of bestiality: power, lawlessness, and otherness. As a consequence, exposure to the sovereign brings forth a mixture of *fascination* at his dual status as human and beast and *fear* of the wanton destruction he is capable of committing. This is fundamentally different to the "shame"[20] that Derrida describes experiencing when naked in front of his cat. It does, however, demonstrate that the human-animal relationship gives rise to different human experiences based on the nature of the interaction and the individual's exposure to and perception of the animal-other.

Bestiality, sovereignty, and criminality are intimately linked to one another and, indeed, can be joined to one another by virtue of their status as that which is beyond-the-law. The difference between them is that the nature of the exclusion from law differs for each: while the beast is always outside the law and so always exists in binary opposition to law, the sovereign starts from a position outside of law to subsequently adopt a dual relationship to law in that he remains external to the law, but yet also exists "within" the law by imbuing it with meaning by virtue of founding and grounding it. In contrast, the criminal only ever becomes one by virtue of breaking the law and so, rather than initially existing outside of the law, is only ever a consequence of a prior law. For this reason, Derrida notes that while beyond-the-law, the beast, the sovereign, and the criminal "have a troubling resemblance" (BS1: 17/38) in that they share "a sort of obscure and fascinating complicity, or even a worrying mutual attraction, a worrying familiarity, an *unheimlich*, uncanny reciprocal haunting" (BS1: 17/38). Derrida is pointing out often-ignored, structural similarities between the beast, the sovereign, and the criminal, all the while resisting the urge to collapse the distinction between them. While sharing a "common" being-beyond-the-law, the "nature" of that "beyond" is different for each, meaning that the three figures cannot be collapsed into one another. They "strangely resemble each other while seeming to be situated at the antipodes, at each other's antipodes" (BS1: 17/38). Despite pointing to certain analogies between the beast and the sovereign, especially in relation to their respective relationships to law, the difference between them must be maintained.

Derrida is not, however, content with pointing out analogies between the beast and the sovereign as if they were simply distinct entities. He makes the stronger claim that the beast and the sovereign are far more intimately and constitutively connected:

> I believe that this resemblance explains and engenders a sort of hypnotic fascination or irresistible hallucination, which makes us see, project, perceive, as in a X-ray, the fact of the beast under the features of the sovereign; or, conversely, if you prefer, it is as though, through the maw of the untamable beast, a figure of the sovereign were to appear. (BS1: 18/39)

It is not the case that the beast and the sovereign are different yet related through analogy. There are logical and experiential relationships between them: when one appears, whether physically or symbolically, so does the other. There are two aspects to this: first, Derrida holds that the phenomenal appearance of each reminds us of the other. Thus, the physical sovereign appears as other than human, a particular beast unconstrained by the norms and rules that define and constrain others. Similarly, the beast, particularly the wolf, appears as something unconstrained by social niceties, norms, or rules. It is, in other words, defined by the ipseity of sovereignty. Again, we see the important role that their shared status as being-outside-the-law plays in their reception. Second, it is not just the physical appearance of each that reminds us of the other, Derrida holds that this is accompanied and strengthened by a symbolic representation: where the physical sovereign appears, he is symbolically associated with a beast, specifically a wolf, marauding through the world. Similarly, the beast is symbolically associated with sovereignty due to his unconstrained engagement with the world. This is not to say that this reminiscence leads one to be replaced by the other. It is simply to say that each appears *through* the other without usurping the other. The beast and the sovereign are, then, in a sense doubled, composed of themselves and the other. This does not, however, introduce a split in each. Derrida is not proposing something as straightforward as a dual-theory of subjectivity or personhood. His point is subtler in that the appearance of each through the other is far more "uncanny" (BS1: 18/39). The physical and symbolic associations manifested through each ensure that the beast and the sovereign appear through one another in a manner akin to a "haunting, or rather the spectacle of spectrality" (BS1: 18/39). Each "inhabit[s] or hous[es] the other [and] the one becom[es] the host of the other" (BS1: 18/39) without this effecting, limiting, or usurping the difference of each. With this, "the beast becomes the sovereign who becomes the beast; there is the beast *and* [*et*] the sovereign (conjunction), but also the beast *is* [*est*] the sovereign, the sovereign [*est*] the beast" (BS1: 18/39).

While this demonstrates the interconnectedness of sovereignty and bestiality, there is something problematic about its continual talk of "the" beast. As Derrida warns in *The Animal That Therefore I am*, we should be wary about talking and writing of

> "The Animal," as if all nonhuman living things could be grouped within the common sense of this "commonplace" the Animal, whatever the abyssal differences and structural limits that separate, in the very essence of their being, all "animals," a name that we would therefore be advised, to begin with, to keep within quotation marks.[21]

Derrida privileges wolf imagery when discussing the beast-sovereign relationship, but his comments on the multiplicity of animality point to the experience of multiple sovereign-beasts. This does not just mean that the sovereign is symbolically linked with different beasts, but that different conceptions of sovereignty entail different configurations of the human-animal relationship: some will appear to be more "human" than "animal," others less so. These topological complications are compounded further by Derrida's insistence that the figure of the beast is never singular: "my animal figures multiply, gain in insistence and visibility, become active, swarm, mobilize and get motivated, move and become moved."[22] "The" beast is never static and singular, but alters and becomes increasingly multiple the more it anchors the notion of sovereignty. What was initially an interesting, if relatively straightforward, questioning of a strict human versus animal opposition gradually morphs into an increasingly complicated questioning of multiple, entwined, moving parts. As one becomes the other, long held supposed oppositions between human and animal, political and nonpolitical, beast and sovereign, male and female, visual and symbolic are complicated to the extent that we get an "onto-zoo-anthropo-theologico-political copulation" (BS1: 18/39).

THE PROPER PLACE OF THE HUMAN

The spectrality of the beast and the sovereign also gives rise to the question of the role and place of the human in this relationship. This is especially so given that sovereignty is always manifested through a particular human and so is always constituted by two bodies: the physical human body of the sovereign and the symbolic, (pre)juridical sovereign body.[23] While Derrida has, up to this point, explained how the sovereign's symbolic, (pre)juridical body is associated with bestiality, there is still the question of how the sovereign's "other" physical human body fits into this. When we examine this tripartite relationship, however, Derrida notes a paradox structured from and around the role and place of the human. After all, the human has long been conceived as being "*on the one hand* superior, in his very sovereignty, to the beast that he masters, enslaves, dominates, domesticates, or kills, so that his sovereignty consists in raising himself above the animal and appropriating it" (BS1: 26/50). But, as noted, the relationship between bestiality and sovereignty also means that "*on the other hand* (contradictorily) [there exists] a figuration of the political man, and especially of the sovereign state *as animality*, or even as bestiality" (BS1: 26/50). There is, then, a juxtaposition between "political man as superior to animality and political man as animality" (BS1: 26/50) that demonstrates and gives rise to "the enigma of the place of man, of what is proper to man" (BS1: 98/142). Such is the importance of this that Derrida even holds that it has "ran through everything" (BS1: 98/142) previously discussed.

The question of the "proper" place of man links *The Beast and the Sovereign* to Derrida's early, influential essay "The Ends of Man" and, in particular, the critique of Heidegger proposed therein. According to Derrida, Heidegger's attempt to reraise the question of the meaning of being and so undermine the pretensions of metaphysical humanism by showing that the human is not the Archimedean point from which all else emanates, ultimately fails because Heidegger's reraising of the question of the meaning of being is always accompanied by and, at least, implicitly dependent on a questioning of "the *proper of man*,"[24] one that acts as a "magnetic attraction"[25] for those, like Heidegger, wishing to overcome anthropocentrism. It is interesting to note that, while, in "The Ends of Man," Derrida criticizes Heidegger's continued magnetic attraction to the question of what is proper to man, *The Beast and the Sovereign* not only once again takes up this issue in the same terms, but also seems to now accept that the question of the proper status of man/the human must be asked and seriously engaged with. While it is tempting to conclude that this simply points to an inconsistency in Derrida's *oeuvre*, another more plausible conclusion would be that, while Derrida shares Heidegger's attempt to overcome anthropocentrism, he is far more reflectively aware that the human cannot simply be surmounted and subsequently left behind, but must be continuously exposed, engaged with, disrupted, and questioned.[26]

For this reason, Derrida's engagement with the question of the proper place of man in the beast-sovereign relationship is far more questioning, hesitant, and open-ended. There is no attempt to provide an axiomatic definition of human being. The beast-sovereign relationship is used to expose what we understand by man/human and, indeed, this entity's relationship to others. It is not the case, therefore, of simply positing the human in opposition to animality/bestiality, but of "wondering whether what one calls man has the right, for his own part, to attribute in all rigor to man, to attribute to himself, then, what he refuses to the animal and whether he ever has a concept of it that is *pure, rigorous, indivisible, as such*" (BS1: 130/182). In turn, the questioning of the proper place/role of the human and his relationship to animals also requires and depends upon an understanding of what is proper to the beast, an inquiry that is, as Derrida realizes, "more difficult still" (BS1: 140/194). The question of what is proper to the beast/human is further complicated because these categories are not only "two plurals (beasts, men)" (BS2: 8/30) but also "asymmetrical" (BS2: 8/30). The beast and the human do not form a perfectly symmetrical relationship; they are not two independent, equal monads facing one another. Their relation is moving, dynamic and so asymmetric. Indeed, it becomes more so once the sovereign's capacity "to-see-without-being-seen" (BS1: 293/391) is taken into account. This allows the human sovereign to see the other, while, at the same time, remaining hidden from its (animal) other. This ties into Derrida's attempt to undermine the dichotomy that sees a privileged human controlling its interaction with a passive animal. In contrast, Derrida emphasizes the ways in which the animal gaze imposes itself onto humans: to Derrida's experience of being naked in front of his cat in *The Animal That Therefore I am* is added the *bestial* sovereign who sees but does not necessarily allow him/itself to be seen. In multiple ways, exposure to animality makes the human strangely vulnerable. But yet, at the same time as this asymmetry is affirmed, it must be remembered that the human sovereign is always intimately connected to the animal/beast, a connection that again calls into question any attempt to establish a straight-forward privileging of the animal over the human or the human over the animal. The animal that imposes itself on the human or that remains at a distance from him also entails the human imposing itself on itself or remaining at a distance from itself. His animal aspect means that there is always an aspect of the human that he cannot perceive. Full self-exposure is not possible; an aspect of himself must remain out of sight.

Rather than provide an answer to the question of the *proper* understanding of the human, animal, or sovereignty or, indeed, the *proper* relationship between them, Derrida is demonstrating the complexity of the issue at hand. It is not possible to simply provide an answer to the question. Indeed, the deferred nature of signification means that a final answer is not possible. We have to learn to properly think by learning to patiently and painstakingly question and engage with much that has previously been taken for granted, specifically the use of clear-cut, nonambiguous, unified concepts and relations. Rather than think in terms of purity and a binary logic of versus, Derrida's exploration of the beast-human-sovereign relationship is undertaken from "a principle of contamination, a law of impurity, a parasitical economy."[27] The aim, an appropriate one for a seminar, but which also appeals to the open-ended nature of Derrida's thinking in general, is to use an issue, in this case the relationship between the beast and the sovereign, to challenge the mode of thinking and, in particular, the binary logic that Derrida claims has defined Western thinking. By exploring the multi-faceted, differentiated ways in which the beast and the sovereign come together, Derrida exposes, interrogates, challenges, and undermines any understanding of this relation that thinks of it in straightforward binary terms. Not only are the human and animal multiple, but so also is the relationship linking them. Exposing the complexity of the bestial sovereign complicates the relationships between "the" masculine and feminine, physical and symbolic, real and imaginary, rational and irrational, and political and nonpolitical. As Derrida puts it, "what the texts we have read call for is at least a greater vigilance as to our irrepressible desire for the threshold, a threshold that *is* a threshold, a single and solid threshold" (BS1: 333–334/442–443). The problem with an unquestioning acceptance of thresholds is that they tend to ossify into fixed boundaries, ones that unjustifiably exclude those deemed to be "other." Rather than continue with this logocentrism, a mode of thinking that has been supported by and, in turn, supportive of, a masculine privileging, Derrida tentatively appeals to a logic based from the feminine hymen to emphasize that while things are different, they are also intimately related and, indeed, permit exchanges. Appealing to the feminine hymen does not entail a mere logical reversal from a privileging of masculinity to femininity, but the attempt to undermine the fixed, impermeable line that supports either privileging. Rather than rely on a logic defined by fixed, closed, monadic terms opposed to one another across a definitive line, *The Beast and the Sovereign* challenges this understanding and, in so doing, dares us "to rethink the very figure of the *threshold* (ground, foundational solidity, limit between inside and outside, inclusion and exclusion, etc.)" (BS1: 333/442). With this, Derrida interrogates the relationship between the beast and the sovereign, the political symbolism inherent to both including our everyday use of animal imagery, and challenges us to think about how we think; a fitting and lasting gift for a final seminar.[28]

NOTES

1. Jacques Derrida, "Force of Law: The 'Mystical Foundation of Authority,'" in: *Acts of Religion*, ed. Gil Anidjar (Routledge: New York, 2010), 231–298.

2. Jacques Derrida, "Before the Law," in: *Acts of Literature*, ed. Derek Attridge (Routledge: New York, 1992),183–220.

3. Jacques Derrida, *The Animal That Therefore I am*, trans. David Wills, edited by Marie-Louise Mallet (Fordham University Press: New York, 2008).

4. Jacques Derrida, *The Beast and the Sovereign, volume one*, trans. Geoffrey Bennington (University of Chicago Press: Chicago, 2009) and *The Beast and the Sovereign, volume two*, trans. Geoffrey Bennington (University of Chicago Press: Chicago, 2011). Citations to these texts will be in-text using the abbreviation BS,

followed by the volume number and the page numbers in the English translation, followed by that of the original French edition.

5. David Farrell-Krell, *Derrida and Our Animal Others: Derrida's Final Seminar, "The Beast and the Sovereign,"* (Indiana University Press: Bloomington, 2013), chapters 1–2.

6. Michael Naas, *The End of the World and Other Teachable Moments: Jacques Derrida's Final Seminar* (Fordham University Press: New York, 2015), 1.

7. Jacques de Ville, "Deconstructing the Leviathan: Derrida's *The Beast and the Sovereign,*" *Societies,* vol.2 (2012): 357–371.

8. Chris Danta, "'Might Sovereignty be Devouring?': Derrida and the Fable," *SubStance*, vol.43, no.2 (2014): 37–49.

9. Mathew Chrulew, "'An Art of Both Caring and Locking Up': Biopolitical Thresholds in the Zoological Garden," *SubStance*, vol.43, no.2 (2014): 124–147.

10. Naas, *The End of the World*, 1.

11. Ibid., 1.

12. Jacques Derrida, *Dissemination*, trans. Barbara Johnson (Continuum: London, 2004), 21.

13. Derrida, "Before the Law," 206.

14. Farrell-Krell, *Derrida and Our Animal Others*, 10–11.

15. Jordan Belford, *The Wolf of Wall Street* (Hodder & Stoughton: London, 2008). In 2013, Paramount Pictures released a movie, directed by Martin Scorcese, of Belford's exploits under the same title as the book and, in so doing, brought wolf-imagery to cinema.

16. Derrida's most extensive engagement with this central concept is found in his "Différance," *Margins of Philosophy*, trans. Alan Bass, (University of Chicago Press: Chicago, 1982), 3–27.

17. Derrida, "Force of Law," 242.

18. Ibid., 244.

19. Paul Patton, "Sovereignty Conditioned and Unconditioned," *SubStance*, vol.43, no.2, (2014): 162–173 (p.164).

20. Derrida, *The Animal That Therefore I am*, 4.

21. Ibid., 34.

22. Ibid., 35.

23. The classic account of the sovereign's two bodies is Ernst Kantorowicz's, *The King's Two Bodies* (Princeton University Press: Princeton, 1997).

24. Jacques Derrida, "The Ends of Man," p.124, in *Margins of Philosophy*, trans. Alan Bass (University of Chicago Press: Chicago, 1982), 111–136.

25. Ibid., 124.

26. For a detailed discussion of Derrida's analysis of Heidegger's anthropocentrism, see my, "Authoritarian and Anthropocentric: Examining Derrida's Critique of Heidegger," *Critical Horizons: A Journal of Philosophy and Social Theory*, vol.16.1 (2015): 27–51.

27. Jacques Derrida, "The Law of Genre," in *Acts of Literature*, ed. Derek Attridge, (Routledge: New York, 1992).

28. This chapter forms part of the activities of the Conex Marie Skłodowska-Curie Research Project "Sovereignty and Law: Between Ethics and Politics"

(2013-00415-026) co-funded by the Universidad Carlos III de Madrid, the European Union's Seventh Framework Programme for Research, Technological Development and Demonstration under Grant Agreement 600371, The Spanish Ministry of the Economy and Competitivity (COFUND2013-40258), The Spanish Ministry for Education, Culture, and Sport (CEI-15-17), and Banco Santander. More information about the project can be found at: https://sovereigntyandlaw.wordpress.com/.

WORKS CITED

Belford. Jordan. *The Wolf of Wall Street*. Hodder & Stoughton: London, 2008.

Chrulew. Mathew. "'An Art of Both Caring and Locking Up': Biopolitical Thresholds in the Zoological Garden," *SubStance*, vol.43, (2014):124–147.

Danta. Chris. "'Might Sovereignty be Devouring?': Derrida and the Fable," *SubStance*, vol.43, n.2 (2014): 37–49.

Derrida. Jacques. "Différance." In *Margins of Philosophy*, trans. Alan Bass, 3–27. University of Chicago Press: Chicago, 1982.

———. "The Ends of Man." In *Margins of Philosophy*, trans. Alan Bass, 111–136. University of Chicago Press: Chicago, 1982.

———. "Before the Law." In *Acts of Literature*, edited by Derek Attridge, 183–220. Routledge: New York, 1992.

———. "The Law of Genre." In *Acts of Literature*, edited by Derek Attridge, 221–252. Routledge: New York, 1992.

———. *Dissemination*, trans. Barbara Johnson. Continuum: London, 2004.

———. *The Animal That Therefore I am*, trans. David Wills, edited by Marie-Louise Mallet, Fordham University Press: New York, 2008.

———. *The Beast and the Sovereign, volume one*, trans. Geoffrey Bennington. University of Chicago Press: Chicago, 2009.

———. "Force of Law: The 'Mystical Foundation of Authority,'" in: *Acts of Religion*, edited by Gil Anidjar, 231–298. Routledge: New York, 2010.

———. *The Beast and the Sovereign, volume two*, trans. Geoffrey Bennington. University of Chicago Press: Chicago, 2011.

De Ville. Jacques. "Deconstructing the Leviathan: Derrida's *The Beast and the Sovereign*," *Societies*, vol.2 (2012): 357–371.

Farrell-Krell. David. *Derrida and Our Animal Others: Derrida's Final Seminar, 'The Beast and the Sovereign.'* Indiana University Press: Bloomington, 2013.

Kantorowicz. Ernst. *The King's Two Bodies*. Princeton University Press: Princeton, 1997.

Naas. Michael. *The End of the World and Other Teachable Moments: Jacques Derrida's Final Seminar*. Fordham University Press: New York, 2015.

Patton. Paul. "Sovereignty Conditioned and Unconditioned," *SubStance*, vol.43, (2014):162–173.

Rae, Gavin. "Authoritarian and Anthropocentric: Examining Derrida's Critique of Heidegger," *Critical Horizons: A Journal of Philosophy and Social Theory*, vol.16, (2015): 27–51.

Chapter 2

The Loaded Cat

David Brooks

Henry Lawson (1867–1922) wrote over two hundred short stories, some of which rank among the best anywhere in the world at the time. And many of the stories involve animals. Perhaps the best known of these are "the Loaded Dog," in which several key characters are canine, "the Drover's Wife," in which, although the story in fact presents us with a panoply of animals, a dog and snake feature, and "the Bush Undertaker," which centers upon a large goanna (and which, in the opinion of Inga Clendinnen, "out-Becketts Beckett at a canter"[1]).

Characteristically, Lawson's employment of animals is more for humor and concerns of the narrative than for a concern with animals themselves, although of course there are further dimensions. Dogs are usually present in their capacity as better or worse friends to man. The snake in "the Drover's Wife" and goanna in "the Bush Undertaker," on the other hand, emblematize a perhaps uniquely Australian threatening (and therefore somewhat Gothic) otherness. There is about them—the goanna especially—something of the uncanny, as if for Lawson an occasional function of the animal is to look at, from outside, and to some extent ironize, human knowing, the eyes of the animal focalizing that manner in which antipodean landscape and experience have been seen to interrogate Western culture.

The goanna is from a family of lizards called *monitor* (in Lawson's story it's very likely a *lace* monitor) supposedly from their capacity to stand on their hind legs and "monitor" their surroundings (*monitor*, from the Latin *monere*, "to warn"). In "the Bush Undertaker" an old bushman—a shepherd—on an expedition to plunder a native grave, comes across the desiccated body of a friend named Brummy, whom he assumes to have died on his way to visit him. He binds the body between sheets of bark (turns it thus into a sort of *tree*) and drags it back to his hut, for "proper" burial.

On the way he notices a large black goanna, and then several more as his journey progresses, and wonders whether there is a sudden infestation of them staring from the bush about him, since he has never seen so many before. As I can testify, having more than once as a child encountered one stretched out on our front fly-screen door, the sight of a goanna, one of the largest lizards, a carnivore, with powerful jaw and claws, can be unsettling. The old bushman is understandably on edge.

At home, he props his friend in a corner and proceeds to comfort himself with drink (readers might suspect a touch of *delirium tremens* in his vision of multiple goannas), but is drawn out into the night by sounds from his roof. It is the goanna—there has only ever been the one—that has followed him because he (or she) has been feeding on Brummy's corpse, and the old man has unknowingly taken off with his/her meal. The goanna throughout the story is *watcher*, and in a sense the old man feels naked before it, perhaps because he is not unaware that there has been something sacrilegious, a profound violation, about his removal of the aboriginal bones. Although to some extent Lawson grimly defuses the situation—the shepherd eventually shoots the goanna—there's a sense in which (*monere*) he has been warned.

There is also the matter of Brummy's name. To be *brum* is to come from Birmingham, a town once infamous for the groats counterfeited there; *brum* therefore coming to mean "false," "counterfeit," "sham." If we think that the goanna is thus feeding on the sham or counterfeit, that the man it's partly consumed is now dressed as a *tree* (our *arborescent* epistemology), and so on, the plot thickens. Whether Lawson *intended* us to read the story in this way—the lizard as monitor/devourer of false knowledge, false philosophy (as, we shall see, Derrida clearly feels his cat's eyes do)—perhaps doesn't matter. The culture writes itself through us.

Given a little metaphoric latitude, there is something of *watching* in "the Loaded Dog" also. The essence of the story is simple. Having at first dragged its fuse through the fire, thereby igniting it, a young and playful dog runs loose with a sausage-shaped—hence its attraction—explosive device in its jaws, mistaking the way his human friends flee from him for an invitation to chase them. One of those friends runs to the local shanty (pub) and the dog follows, sending the drinkers scurrying. Two of his other friends flee to a second building, the shanty's kitchen. Trying to get to them the dog runs underneath it only to be chased out again and have his "sausage" taken by "a vicious yellow mongrel cattle-dog" that has been "nursing his nastiness" under the floor. The mean dog settles down with his prize, observed by a dozen others who have suddenly materialized—"spidery, thievish, cold-blooded, kangaroo-dogs, mongrel sheep- and cattle-dogs, vicious black and yellow dogs—that slip after you in the dark, nip your heels, and vanish without explaining—and yapping, yelping small fry"[2]—and of course, the device at last exploding, the "vicious yellow mongrel" is no more.

The men the dog runs among are gold prospectors, and while it may be a little fanciful (but texts *do* run away from us) to invoke David Hume's discussion of a *golden mountain* in his *An Enquiry Concerning Human Understanding* (1748), or to ask the reader to envision these prospectors as philosophers (or philosoph*ies*), the dog does have, *in his mouth* ("And say," asks Derrida, "the animal responded?"), the capacity to *explode* his human companions, and the configuration—the underlying ecology—is (implicitly) there. Fanciful, as I said, but indulge me a while.

Although it may interest the reader who does not already know the story to hear that the aforementioned explosive device was originally fashioned in order to harvest fish from a water-hole, and that when it eventually explodes, killing the "yellow mongrel" (can we divorce this "yellow" from the racist xenophobia so endemic in Australia at the time, especially on the goldfields?), a number of the other dogs are also maimed and traumatized, and although one cannot help but reflect that if the dogs here so unflatteringly listed are unattractive and ill-tempered it is very likely, since so many of them are "working" dogs, that it is humans who have made them so, there is perhaps no need to discuss this story further (at least, not just yet). *This* essay, after all, is called "the Loaded *Cat*," and I have introduced Lawson's story to some extent as a means of *loading* a brief discussion of Derrida.

But first there is "the Drover's Wife," a story about a woman, a slab-and-bark hut—or rather the large, earthen-floored kitchen at the end of it—and a snake.

The short third paragraph of this story presents the situation concisely: "the drover, an ex-squatter, is away with sheep. His wife and children are left here alone." Just so. And the story proper—its central action—begins with the sighting of a snake—a very poisonous snake—that disappears first into the woodpile, and then under the house.

It is almost evening. The drover's wife—she's not given a name—ushers her children into the kitchen, feeds and puts them to bed, and, with the dog (named "Alligator"), stays up all night watching. As she sits awake she goes over in her mind some of the previous trials she has experienced in the Australian bush alone: a bushfire, a flood, a mad bullock, the illness and death of children, threatening and insistent swagmen nosing at her door. "An evil pair of small, bright bead-like eyes"(*monere*) eventually appears at a hole in the wall between the house and kitchen, and, although sensing danger and attempting immediately to retreat through a neighboring hole, the snake at last enters, to be caught and killed by the woman and her dog, working together.

The story then ends, melodramatically, with the elder son, Tommy, declaring to his mother, as they watch the "mangled reptile" burn in the fireplace: "Mother, I won't never go drovin'; blast me if I do!" From a sheep's perspective, it would be nice to think that this registered an animal turn, but of course it is nothing of the kind. To suggest that the story is entirely unambivalent in this regard, however, would be wrong. Earlier in the story, as the children settle down to sleep on the kitchen table, Tommy is given to say "Mother! Do you think they'll ever extricate the (adjective) kangaroo?" Idle chatter, it may be, but, from the author of "The Bush Undertaker," we might be inclined to ask why the eradication of the kangaroo should be mentioned at all, if we are not being asked to reflect upon it, and perhaps to look at the malapropism more closely. "Extricate"? "to free someone or some thing from a constraint or difficulty"?[3] Within the story, that is to say, there could be seen to be something like a hole, a trapdoor, into some other space entirely.

Whether this kind of hole is in any way analogous to that "horrid black hole" so central to D. H. Lawrence's remarkable "snake" is another matter. Certainly Derrida, who discusses this poem in the ninth of his seminars in *The Beast and the Sovereign*, makes no mention of it. In the poem, Lawrence's *persona* comes out of his house in Sicily on a hot day to get water, to find a snake at his water trough. The vision captivates him. It's a moment of deep encounter. He is moved, honored, that so regal a creature has come to share his water. And torn, since as a human male he has been conditioned to think that he should be killing, not admiring. But he restrains this impulse. Gazes instead in wonder. Until the snake, having drunk, begins to retreat into the hole in the wall from which it came. Something then erupts in the watcher, a kind of fury—at the thought that the snake is leaving, perhaps, or that it is entering, through its "horrid hole," the wall of his own house. Putting down his pitcher (sign of openness, of receiving), he then, cum the drover's wife, picks up a lump of wood, and throws it at the snake.

What *is* this hole? What is this horror? It's as if the snake's withdrawal, its self-sufficiency, its disinclination to share the human space and moment, exposed some rupture, some deep *in*sufficiency in the persona's very being, the mere existence of the hole—access, potentially, to *another place*—giving the lie *to* that being. Derrida reads the poem on a double axis of host and guest, beast and sovereign, and although I think his reading thin and erroneous, perhaps there is something there too, in the persona's assumptions of sovereignty—such as we can see in the complacent/superior *position* of the gaze (admiration extended like a gift, a raising-up), the trimmings of ownership ("my water-trough," "my wall-face," even "my snake")—that are in effect revealed to be illusion, imposture, have no *grounding*, by the snake's disregard and returning to darkness, so that the black hole (prelude to Derrida's abyss) becomes a kind of wound, a site or symbol of castration (the cat in Derrida's bathroom, we will find, in the moments before she demands egress, stares at the philosopher's *penis*).

It is time, that is to say, that we turned to Derrida, or rather—a digression that is also a gathering—to one of his titles, and then to that now-famous encounter with his cat.

SUIVRE

Derrida's *The Animal That Therefore I Am* was published in France in 2006 and in English translation in 2008,[4] though these dates are of course a little deceptive. The book is in fact a transcription, and then translation, of notes from some long lectures he gave ten years before to a conference on his work in Cerisy-la-Salle. The conference was entitled "the Autobiographical Animal" (*l'Animal autobiographique*), and the first lecture was itself entitled "The Animal That Therefore I Am."

Clearly this title (a translation of *L'animal que donc je suis*), recalls directly Descartes' *cogito ergo sum*, "I think, therefore I am" (*Je pense, donc je suis*). Clearly too, any recollection of Descartes in writing about the Animal is doubly loaded. He may have given us the *cogito*, but he made, also, the infamous claim that animals are mere machines—*automata*—incapable of conscious thought and human-like feeling. Descartes' position is more nuanced than commonly recognized and I would not want to simplify it, but his derogation of animal sentience is emphatic, and, for those who venture beyond the title, a good deal of Derrida's ensuing address is devoted to an implicit critique of this aspect of Descartes.

Perhaps the most obvious implication of Derrida's recasting of the Descartes is the positive negative: "No, it is not that you Are because you Think, but because you are *animal*." But there is also ambivalence. On the one hand, the title can be read as "because of certain arguments [to which this title refers with its 'therefore'], I am an animal," and on the other as "I am, *because of* the animal." Given that within the lecture itself Derrida criticizes (and attempts to replace with the awkward term *animot*) the word "animal" as an umbrella term—the word "animal," he says, is an intellectual *violence*, a means of ignoring the *species particular* and *individuality* of the creature(s) we pretend to talk about—this second reading of this title can also be seen to refer to the manner in which we employ such an umbrella term, that is, speak of "the Animal," as a means of maintaining our distance and separation *from* it: *abjecting* it, as it were, in the process of defining ourselves, casting *from* ourselves what we construe to be "the Animal" in us. We *invent* "the Animal," which does not exist (there exist only dog, cat, badger, kangaroo [but these too are umbrella terms!]), in order to emphasize our understanding that this (the Animal) is what we are *not.*

Leaving aside here the amusing irony that we thus define ourselves by a kind of double negative, that is, by holding that we are not something that, as it happens, *is* not (does not exist), we might bear in mind that, were we to accept Derrida's point that, in speaking of "the animal," we should in fact *list specifically* those animals of which we speak, to speak in effect of "the Animal" as we might think we understand it we must at every point list so many specific creatures that the vastness of the task will almost certainly reduce us to silence, and so, rather than closing the gap between human and nonhuman animals, as he is generally taken to be attempting to do, Derrida might in effect be only widening it.

We might also bear in mind (a point he himself seems to accept, in the book of this title, with some embarrassment and without explanation) that Derrida's demand for specificity here is in effect an assertion of a fixedness and *real*ism in language quite in contravention of the inherent *in*specificity he has spent his career demonstrating. (Thirty years or so of reading Derrida, on and off, and I have come to the conclusion that he is like this; that he gives but very often at the same time takes away; that he can be radically subversive in order to be radically conservative.)

But there is more. "I think, therefore I am"/"the animal [that] therefore I am." What is the force of this substitution, that is, of "the animal" for thought? Which is the primary term? Are we meant to understand that this substitution is possible *because* animals think? Or are we meant to understand it as a *correction* asserting (as we have seen) that one (a "human") *is*, *not* because one *thinks*, but because one is *animal* (and potentially, thereby or therein a statement that *de-centers* the issue of thought, removes or challenges it *as* an issue)? Or is it, quite differently again, that, through this substitution, there is a *negation* of thought, that is, a suggestion or implication that animals *do not* think, or that the animal is something *other than* thought?

Whatever else it does, this substitution of "animal" for "think" serves to maintain these terms in relationship, holds them in one another's orbit at the same time as it keeps them apart, as if to acknowledge—a further possibility—that thought can only ever approach the animal *through* (thought) *itself*, but is aware at the same time that what it pursues cannot be approached in this way.

There is also—a further and very significant oscillation—the verb *suivre* ('to follow"). Different as the verbs *être* and *suivre* might otherwise be, their first person present forms are identical. *Je suis* means both "I am" and "I follow." "I think, therefore I follow"? In his essay Derrida refers to this repeatedly, indeed could almost be said to be haunted by it. It might be objected that in making so much of this Derrida allows himself to be distracted by a *gallicism*, by which I mean only that, in English (or German, Mandarin, Russian) the first person singular of the verb "to follow" is *not* the same as the first person singular of the verb "to be," and that what Derrida notices, and is unquestionably *universalizing*, is just as likely to be an accident as it is to be a deep perception, on the part of the French language, of the nature of all human Being. But perhaps this does not matter. If it is an accident, Derrida might argue, it accidentally recognizes something that does in fact exist.

The impact that this has on our reading of his title—this thread that binds his entire essay/lecture together—is manifold. The possibility that to *be* is somehow to follow, is now always with us. But Derrida has of course already complicated this, given that in his title he has substituted "the Animal" for "I think," so that in effect the ghosting by the verb *to follow* is a double one: we have not only "I think therefore I follow," but also "*the animal* that therefore I follow." We must not only try to wrap our minds about each one of these, but also about their interrelation. We must consider not only that *thought—thinking*—entails—*is*—somehow a *following*, but also consider in what way "the Animal" and "to follow" interrelate. Is it the same thing to say that, in order to *be*, one somehow *follows the animal* as it is to say that, *because* of the animal, one (somehow) follows—without, one needs to say of the latter interpretation, any clear indication of what it is that one follows?

One also has to ask what possible use or benefit there might be to animals in following through such equivocations, but I would ask my reader to set that aside for the moment. There is a further matter to be considered. It may be just one more distraction. It may be, on the other hand, that it helps to lead us out of what must be coming to seem a bit of a maze.

Derrida's book is, as I have already indicated, a transcription—in English a *translation* of a transcription—of a lecture given in 1997. It is possible—Derrida is often a step ahead of us—that the lecture remained to be brought to us in this round-about way quite deliberately. It does, after all, take us back to one of the roots of philosophy, the lectures—dialogues—which Socrates and his collocutors delivered verbally and left to Plato to transcribe, so that the written arguments are always haunted by the spoken. It takes us back, too, to Derrida's own lengthy deliberations about this very point in *Of Grammatology* and elsewhere. Indeed his early reputation as a philosopher rests upon the extrapolations he made upon this phenomenon, that is, of there being always a text before the text, no "end" to writing: within the text of Plato there is the "text" of Socrates; within the text of Socrates there is the text of the voice within Socrates, a version of which—it can never be more than that, an "edited" version—Socrates delivers in his public pronouncements. And within or behind the text-within there are the texts that have nurtured and enabled it—the past of the language, the past of thought, past philosophers, past conversations, and so on. And every one level of this text—text*ure*—is an edited version of its prior text/s, filled/haunted by the *traces* of what has not made it through.

And we might at this point note the kinship between the verb *to trace* and the verb *to follow*. We might also note that for Derrida there is, for humans, no being without text, that in a sense *to be* is to be text, that there is, for us, no being that is beyond or outside text.

It may be, that is to say, that Derrida has *deliberately* left his lecture—his own involvement with his lecture—in oral form, to emphasize at once its own aurality, and its deep relation to this long current of his thinking. The verb *suivre* is perhaps a symptom of this. To an audience *listening*, particularly once he has introduced and drawn attention to the pun, the words *je suis*, whenever they are uttered by the speaker, will have this doubleness. I don't dispute that it will not be there in the French transcription also, but it will have lost some of its force, albeit in the gaining of a different kind of force: its effect will now be all the more conscious and conceptual, an intellectual problem, than a haunting. And, of course, in the English (or German, or Mandarin) it will not be there at all. All these readers will have will be Derrida's insistence that it exists. Its trace will be there but significantly thinned, muted, attenuated.

And there is *animot*, a combination of the word for *animal* and the word for *word*. A term which Derrida offers us as a means of emphasizing, and remembering, his point that there is no such thing as *the animal*, that it is an intellectual violence, a lump-word, an umbrella term, a sleight of language, to keep animals at bay. (A tautology, there, in my mode of expression, but then I have written already about the problems of this argument of Derrida's, the impotence that it brings about, the dangerous silence that it threatens to impose.) Which is only, at this point, to say that whenever Derrida, in his lecture, uses the word *animaux*, it will be ghosted—because it sounds exactly the same—by the word *animot*, and vice versa. Something, a haunting, which is lost, clearly differentiated, in the written version. (As perhaps is a further haunting, almost unbelievably *not* pointed by Derrida, but then it may be that it is just too obvious: that a further fracture of "animal" is into "ani-" and "*mal*"; that within the word is a sickness, a *malaise*, carried over into the plural, "ani*maux*," so that "ani*maux*'/'ani*mot*," word and sickness are conflated, overlaid.)

What is all this to say then? Perhaps only that, locked within language, a being that *is* because it is text, a being for whom textness and being are inseparable, the human, because text is always preceded by text—because in this way there is no end to writing—is a creature divided, doubled, always itself but that self always a self that follows, that shadows, that is produced by prior writing; a creature that cannot escape language and the thought that—while it is certainly possible to think without language—cannot be apprised, realized, communicated, without it. So that (a) human being is a being *within* being, held somehow apart from a greater being by the very language it uses in order to apprehend and apprise its being. And that, perhaps, not just through the particular (*animaux*, *animot*), we have separated ourselves from the animal through language, or rather that it is language which has separated us from it/them. That *we* are what we are because of the ani*mot*. That we have used the umbrella term to separate us from the animal particular.

Ultimately, perhaps, we are not so much to make a choice as we are to see these two major *fields* of interpretation that a close consideration of Derrida's title seems to open before us—the field, on the one hand, of our entrapment within language and of our concomitant linguistic *abjection* of the animal as part of the means by which we have defined ourselves as human, and on the other hand, the field of a greater being into which we might emerge/escape, were we able in some way to follow the animal—as existing in oscillation. *Je pense donc je suis*: the statement and its manifold traces. What it is and what, in order to be, it has shut out but which it can never utterly repress. We are separated, by language, from the animal; by language we are also separated from ourselves; we are not integral; being is also a following; we need the animals to be; we may follow animals into being.

Now to meet the cat herself.

THE CAT

She appears almost immediately, as if Derrida has been keen to tell us about her. Sitting in his bathroom—locus of abjection, site of preparing-oneself-for-the-world—observing the philosopher naked. Which makes the philosopher, *as* philosopher, *feel* naked, an experience not unlike that of the Emperor in Hans Christian Andersen's "the Emperor's New Clothes," when the young boy in the crowd, uninhibited by the fear and self-doubt silencing the people around him, observes loudly that the Emperor before them is clothed in nothing at all. Before the gaze of the animal—such is Derrida's assertion—the philosopher is embarrassed, his philosophizing no longer able to clothe him.

Embarrassed, that is, and also somewhat apprehensive, since we are given quite clearly to understand that the focus of the cat's attention is Derrida's penis. A *castration anxiety* is therefore involved, and were the author anyone else but Derrida we might think that that was all it was and, perhaps a little abashed that this information has been shared with us, move on. But of course this goes to the core of the matter. We are talking—*Derrida* is talking—obliquely, of phallogocentricity, and the challenge, the threat of castration, that the Animal presents to it.

But how appropriate is Derrida's choice of the cat as the vehicle of this perception? A creature—a species—that has not only accompanied the human, has been *domesticated*, for thousands of years, but that has such a presence in folklore, mythology, literature, even philosophy, that it is hard to see how it can be in any way an innocent player. The cat, we could say at the outset, is *loaded*, so heavily in this way—Derrida has even (although he does not tell us this) named her "Logos"[5]—that one wonders whether she can, in truth, be seen as anything other than *mirror*. She may, that is, sit Sphinx-like observing him, her stance unclothing his philosophy, but (and I must add that Derrida himself nowhere mentions the Sphinx) she *is* Sphinx-like—few readers will not find such a thought passing through their minds (i.e., *themselves* loading the cat)—and the Sphinx, we might reflect, was not only half-lion/half-human (already nonhuman animal and not), but presented Oedipus most famously with a riddle (what goes on four legs, then on two legs, then on three?) the answer to which was Man.

I would wonder, too,—Derrida is Francophone after all—if there wasn't, in any writing he might do about a cat, more than the ghost of a memory of Baudelaire's famous poems, "Le Chat" and "Les Chats," the latter of which was the subject of one of the key essays (by Roman Jakobson and Claude Lévi-Strauss, published in 1962) in Structuralist analysis.[6] And if it is hard to see how Derrida could write about a cat without the Sphinx, or Baudelaire, or Jakobson and Levi-Strauss in mind, it is almost impossible to imagine that he could do so without some thought to Maurice Blanchot. Blanchot's early uses of the cat—or, rather, the texts in which those uses occur, but one must emphasize that these uses are central to these texts, as well as to the definition of Blanchot's own relation to the Animal—are of considerable significance to the subsequent development of literary theory and its intensification of the interrelation of literature and philosophy. Blanchot himself was a major influence in the development of Derrida's thought, was eventually a friend (of a sort) of Derrida, and was the subject, indeed nominal co-author, of one of the last books published by him.[7]

'Everyday language," writes Blanchot in his signal 1947 essay "Literature and the Right to Death":

> calls a cat a cat, as if the living cat and its name were identical, as if it were not true that when we name the cat, we retain nothing of it but its absence, what it is not. Yet for a moment everyday language is right, in that even if the word excludes the existence of what it designates, it still refers to it through the thing's nonexistence, which has become its essence. To name the cat is, if you like, to make it into a non-cat, a cat that has ceased to exist, has ceased to be a living cat, but this does not mean that one is making it into a dog, or even a non-dog[8]

—lines which could be seen as themselves an explanation of a pivotal passage in the revised version of his first novel, *Thomas the Obscure*:

> Thomas got up and went silently downstairs. No one noticed him with the exception of a nearly blind cat who, seeing the night change shape, ran after this new night which he did not see. After slipping into a tunnel where he did not recognise a single smell, this cat [...] filled his lungs and howled. He drew from the idol he was becoming the incomprehensible voice which addressed itself to the night and spoke.

"What is happening?" said the voice. "the spirits with which I am usually in communication, the spirit that tugs at my tail when the bowl is full, the spirit that gets me up in the morning and puts me to bed in a soft comforter, and the most beautiful spirit of all, the one that meows and purrs and resembles me so closely...

have all disappeared. Where am I now? If I feel gently with my paw, I find nothing. [...] I am surrounded by a special void which repels me and which I wouldn't know how to cross over. [...] I have a hideous affliction: my face feels as large as a spirit's face, with a smooth, insipid tongue, a blind man's tongue, a deformed nose incapable of prophecy, enormous eyes without that straight flame which permits us to see things in ourselves. [...] I am already darker than the shadows. I am the night of night. Through the shadows from which I am distinguished because I am their shadow, I go to meet the overcat. There is no fear in me now. My body, which is just like the body of a man, the body of the blessed, has kept its dimensions, but my head is enormous [...] and rather than a head seems nothing but a glance, just what is it? I can't look at it without uneasiness. It's moving. It's coming closer. It is turned directly toward me and, pure glance though it is, it gives me the terrible impression that it doesn't see me. [...] I am dead, dead. This head, my head, no longer even sees me, because I am annihilated. For it is I looking at myself and not perceiving myself."[9]

This cat, quite simply, has strayed into a novel; strayed, or been drawn, into *language*, into *text*. Deprived thus of his *Dasein*, his "actual" being, *in the realm of the human*, he finds himself at once in a peculiar *emptiness*, and taking on (Sphinx-like) *human* form, since the form of cat *per se* is impossible, is annihilated in this place.

One wonders whether, given a chance to respond, *allowed to get a word in edgewise*, in that *impossible* place of response, Derrida's cat might say something similar. That is, that in Derrida's text she is not *seen*, has taken on *human* form, has become, as it were, a mirror. We should perhaps try to leave aside for the moment the intriguing question of what it might mean for Derrida, master of the trace, of *différance*, he who says that there is *nothing that is not text*, to try to remove this cat from such a situation by proclaiming her *reality*—although we should at the same time recognize that it is a paramount question. Indeed Derrida himself seems somewhat shocked, as if realizing that he has just attempted to strip the cat naked of all association, *to draw her out of the cloud of language*, against so much that he has argued before.

Derrida himself, to acknowledge this at the outset, seems conscious of this seeming paradox, and of the loading which attends it, and tries, in the very act of introducing his cat, to fend some of it off. "I must make it clear from the start," he says,

> the cat I am talking about is a *real* cat, truly, believe me, *a little cat*. It isn't the *figure* of a cat. It doesn't silently enter the room as an allegory for all the cats on the earth, the felines that traverse myths and religions, literature and fables. [374/*6*]

Indeed he seems so nervous on this point (or is he merely tongue-in-cheek?) that he repeats it in the next paragraph. "An animal looks at me.

What should I think of this sentence? The cat that looks at me naked and that is *truly a little cat, this* cat which I am talking about, which is also a female, isn't Montaigne's cat":

> Nor does the cat that looks at me naked, she and no other, the one I am talking about here, belong, although we are getting warmer, to Baudelaire's family of cats, or Rilke's, or Buber's. Literally speaking at least, these poets' and philosophers' cats don't speak. "My" pussycat (but a pussycat never belongs) is not even the one who speaks in *Alice in Wonderland*. Of course, if you insist at all costs on suspecting me of perversity—always a possibility—you are free to understand or receive the emphasis I just made regarding "really a little cat" as a quote from chapter 11 of *Through the Looking Glass*. Entitled "Waking," this penultimate chapter consists of a single sentence: "and it really was a kitten, after all" [...]
>
> Although time prevents it, I would of course have liked to inscribe my whole talk within a reading of Lewis Carroll. In fact you can't be certain that I am not doing that, for better or for worse, silently, unconsciously, or without your knowing. [376/7]

Ah, so there we have it, not only an admission that the cats of Montaigne, Baudelaire, Rilke, Buber, and doubtless many others (Blanchot, Céline) *have* been in his mind, but that one cat, a kitten in fact, from *Through the Looking Glass*, has been especially so. "I would of course have liked to inscribe my whole talk within a reading of Lewis Carroll. In fact you can't be certain that I am not doing that." What to make of this? An admission that the mirror *is* a problem? That, ideally, somehow, one should be able to pass *through* that mirror? An admission, since this *is* a reference to Alice's *waking*, that the texted/textual world/prison *is* like a dream, from which there might, just, somehow, *be* a waking, into a world of "reality"(Rilke's, Heidegger's "Open," let's say)? Perhaps, though one must also remember that Alice woke in her own armchair (holding the black kitten), on the original side of the mirror, with, or as, an implication that she had never passed through at all.

The trembling that we see here, the almost-admission of a need or desire to grasp a *real*, a mode of Being unshackled by a consciousness of its relentless textuality, characterizes much of *The Animal That Therefore I Am*. Derrida asserts that the cat of whom he speaks is herself alone, that she is a "*real*" cat, that she is not "the figure of a cat," and then admits that even the statement that she is just a kitten is a quotation, as if the whole has been a game, a gambit, to demonstrate the point that he has been laboring for years, that there is nothing that is not text. But the mere fact that he would exercise this gambit, play this game at all, seems to betoken a discomfort, a heightened consciousness—before the stare of this cat—of a constraint, an inadequacy, an impairment in this very position. When later in the essay he speaks very directly about the horrors of factory farming and our industrialization of the animal, a process he readily compares to the *Shoah*, it is all too briefly, and is swiftly followed—as if he has shocked even himself and is speaking in a mode he simply cannot sustain—by a return to more guarded, "philosophical" discussion.[10] There is deferment. We are promised an eventual and extensive examination of philosophy's relation to the animal. And when this at last comes, in the form of *The Beast and the Sovereign*,[11] it is itself a kind of deferment.

What has happened? Has Derrida reached, embarrassed, the limits of (his) philosophy? Has he offered himself up as an example of its failure? Has he *stumbled*, in the face of the animal? Has he opened this door at last, this great door, just a crack, and looked through, to find the territory so vast, so strange, that he feels naked before it, and has had to close it and back away?

I suspect that it is all of these things, if these things are separate at all. Or rather, to put this in another way—and come to a rapid and abrupt conclusion—that Derrida has realized that, behind the mirror-cat, which he, and a long tradition behind him, has so loaded that he cannot see past her, there is another cat, another *being*, loaded with herself, her suffering, the weight and intensity of her own existence. An *abyss*, in short, which threatens to swallow him.

NOTES

1. "About Bones,"*Agamemnon's Kiss* (Melbourne: Text Publishing, 2007), 78.
2. My copy-text for these stories of Lawson is *Henry Lawson: Short Stories and Sketches, 1888–1922*, ed. Colin Roderick (Sydney: Angus & Robertson, 1972).
3. However else we might see it, the kangaroo is certainly enmeshed, entangled disastrously in the Western mind. But that, perhaps, is for another essay. Did Lawson "see" animals? That's hard to say. His keen sensitivity toward the sufferings of the working class, especially when pitted (an animal metaphor?) against what was so often seen as the "harshness" of Australian circumstance, sometimes extended to non-human animals, but by and large they are foils for the human animals his narratives follow. Each of the stories discussed could be said to occur within margins of (animal) abuse. The dog is loaded with an explosive intended to kill fish; the explosion which kills the (abused) "yellow" dog traumatises others; the drover is away with sheep, most probably—since there is drought—to take them for slaughter; the bush undertaker neglects his flock and ultimately shoots the goanna. Whether Lawson intends us to see this, rather than simply to present his world realistically, is another matter. Most commentators have paid it no attention.
4. Paris: Galilée, 2006 and (trans. David Wills) New York: Fordham University Press, 2008, respectively. Quotations in this essay are from the original translation of chapter 1: "The Animal That Therefore I Am," trans. David Wills, *Critical Inquiry* 28.2 (Winter, 2002): 369–418.

5. Elaboration upon which I will resist, tempting as it is to suggest that Logos is thus already externalised, ceded to the animal, almost to the point where philosophy, in the form of the philosopher, is placed *outside* the Logos (as distinct from philosophy's more normative positioning of the animal as the Outside *of* the Logos).

6. Available variously, for example in Roman Jakobson, *Language in Literature*, trans. Krystyna Pomorska and Stephen Rudy (Cambridge, MA.: Harvard University Press, 1987), 180–97.

7. Maurice Blanchot/Jacques Derrida, *The Moment of My Death/Demeure*, trans. Elisabeth Rottenberg (Stanford, CA: Stanford University Press, 2000). Derrida's text was originally published as *Demeure: Maurice Blanchot* (Paris: Galilée, 1998). Kevin Hart, in personal correspondence, cautions that to present the two works together was a publisher's decision and should not be taken to evidence any particular personal relation between the authors.

8. "Literature and the Right to Death" (1947), in *La part du feu* (1949), trans. Charlotte Mandell as *The Work of Fire* (Stanford, CA: Stanford University Press, 1995), 325.

9. It is generally held that Blanchot wrote *Thomas l'Oscure*, originally a book of over two hundred pages, between 1932 and 1940. First published in 1941, it was in the late 1940s extensively revised and shortened to approximately one-third of its original length. Published in 1950, this revised version was translated by Robert Lamberton and published in English in 1973. My quotation here is from *The Station Hill Blanchot Reader*, ed. George Quasha (Barrytown, NY: Station Hill Press, 1999), in which Lamberton's translation is reproduced on pages 51–128.

10. Donna Haraway writes of Derrida's here failing "a simple obligation of companion species":

> he did not become curious about what the cat might actually be doing, feeling, thinking, or perhaps making available to him in looking back at him … Incurious, he missed a possible invitation, a possible introduction to other-worlding. Or, if he was curious when he first really noticed his cat looking at him … he arrested that lure … with the sort of critical gesture that he would never have allowed to stop him in his canonical philosophical reading.

——— *When Species Meet* (Minneapolis, MN: University of Minnesota Press, 2008), 20.

11. Edited by Michel Lisse, Marie-Louise Mallet and Ginette Michaud. First published as *Séminaire: La bête et le souverain* by Éditions Galilée in 2008 and 2010. English translation by Geoffrey Bennington (Chicago, IL: University of Chicago Press, 2009 [vol. I] and 2011 [vol. II]).

WORKS CITED

Blanchot, Maurice. "Literature and the Right to Death" (1947). *The Work of Fire*. Translated by Charlotte Mandell. Stanford, CA: Stanford University Press, 1995.

———. *Thomas the Obscure*. Translated by Robert Lamberton. In *The Station Hill Blanchot Reader*, edited by George Quasha, 51–128. Barrytown, NY: Station Hill Press, 1999.

Blanchot, Maurice, and Jacques Derrida. *The Moment of My Death/Demeure*. Translated by Elisabeth Rottenberg. Stanford, CA: Stanford University Press, 2000.

Clendinnen, Inga. "About Bones." *Agamemnon's Kiss*. Melbourne: Text Publishing, 2007.

Derrida, Jacques. "The Animal That Therefore I Am." Translated by David Wills. *Critical Inquiry* 28.2 (Winter, 2002): 369–418.

———. *The Beast and The Sovereign*. Edited by Michel Lisse, Marie-Louise Mallet and Ginette Michaud. Translated by Geoffrey Bennington. Chicago, IL: University of Chicago Press, 2009 (vol. I) and 2011 (vol. II).

Haraway, Donna. *When Species Meet*. Minneapolis, MN: University of Minnesota Press, 2008.

Jakobson, Roman, and Claude Levi-Strauss. "Baudelaire's 'Les Chats.'" In *Language in Literature*, edited by Krystyna Pomorska and Stephen Rudy, 180–97. Cambridge, MA: Harvard University Press, 1987.

Lawson, Henry. *Short Stories and Sketches, 1888–1922*. Edited by Colin Roderick. Sydney: Angus & Robertson, 1972.

Part II

POSTHUMOUS ENCOUNTERS

Chapter 3

"The Most Famous *Dog* in History"

Mourning the Animot in Abadzis' Laika

José Alaniz

Laika, a Moscow mongrel stray, the first terrestrial being to live—and soon after, die—in space, achieved her trailblazing feat on November 3, 1957, orbiting the Earth aboard the Soviet spacecraft *Sputnik-2*. It proved one of the galvanizing events of the Cold War. But as contemporaries suspected, Sovietologists argued and Russian scientists fully acknowledged only in 2002,[1] Laika's story held even more tragedy and less scientific utility than advertised—it amounted to little more than a Khrushchevian propaganda coup bought with a helpless animal's life.

For unlike other canine test subjects trained by the Soviet Union, Laika was sent into space with no intention of bringing her back; moreover, she was not "painlessly euthanized"[2] a few days into the journey, but rather experienced an agonizing death through heat and stress within five hours of launch.[3] Even Dr. Oleg Gazenko, of the Soviet institute which was in charge of Laika, admitted in 1998: "The more time passes, the more I'm sorry about it. We shouldn't have done it. We did not learn enough from the mission to justify the death of the dog."[4]

Indeed, Laika discourses up to the present day have consistently partaken of, even wallowed in, the language of regret, melancholy, repentance—what Amy Nelson calls "associations with sacrifice, loss, exploitation and experimentation" which "fueled" the dog's posthumous celebrity.[5] Since her doomed mission, Laika has become an icon of Russian and world popular culture, taken up by those who both celebrate the USSR's space program (Soviet propaganda) and vilify it (e.g., animal rights groups); manifesting in the mutt's ceaselessly reproduced image, its iterations catalogued by Olessya Turkina *et al.,* in their 2014 book *Soviet Space Dogs:* confectionary tins, cigarette boxes, postcards, cartoons, Argentine comic books, Japanese spinning tops, the Swedish film *My Life as a Dog;* and crowned by public remembrances, from the first memorial, in Paris, dedicated to her and other canine astronauts (1958), to Pavel Medvedev's 2008 *Monument to Laika* in Moscow, erected on the occasion of her flight's 50th anniversary; to "an astonishing number" of websites, songs, music groups, and poetry, as set down by Nelson.[6] In all these expressions, just below the surface (and to varying degrees of denial), lies interred a lament for Laika's sad fate. That Parisian memorial states, in part: "died without burial, martyrs for science."[7]

In 2007, British comics artist Nick Abadzis[8] took the broad outlines of the beloved Russian-stray-turned-space traveler's story, added much background on the USSR's space program at the time, and fictionalized some details to produce his acclaimed and deeply moving graphic novel *Laika.*[9] Central to his version: an explicit, unsentimentalized adherence to the facts of Laika's untimely death—which most other representations either gloss over in maudlin reverence or, as noted, keep under the surface.

As it turned out, for some the novel adhered to the facts a bit too closely—prompting an outcry from readers over the "heartbreaking" conclusion and demands for a "happy" ending. The artist's response to these pleas, explored further on, present a remarkable example of what James Stanescu, writing after Judith Butler, terms the "oddly fecund nature of loss," that quality of mourning which "does not have to stay oriented toward the past, but can instead reorient itself to our present in order to reorient our future."[10]

This essay delves into Abadzis' *Laika*—both the graphic novel and its "afterlives"—as several things: a touchstone in comics' representation of animal subjectivity; a visual-verbal exploration of cross-species affectual encounter; and in particular as a startling instance of twenty-first-century grief. Among the questions I will touch on: what capacities does graphic narrative bring to the depiction of animal subjectivity, animal death, and its commemoration? How does this novel play into what Turkina *et al.* call the "overwhelming global admiration" of Laika? How do Derrida's insights into seeing (and not seeing) shed light on Laika and her representations? And how does readers' refusal to accept the death of their nonhuman heroine comment on modern modes of mourning (its "work" or "output"); on fantasy and death denial in the digital age; and on the cross-species ties forged—and betrayed—in Abadzis' depiction of what one reviewer dubbed "the most expensive and premeditated execution of an animal that the world has ever known"?[11]

ANIMALS IN COMICS

The depiction of animals in graphic narrative has long exhibited the paradoxical visual ubiquity and marginalization-as-metaphor which John Berger regards a hallmark of modernity. Of especial relevance: his discussion in his 1980 essay, "Why Look at Animals?" of the nineteenth-century French caricaturist J. J. Grandville's work, whereby animals *qua* animals all but disappear in a pictorial strategy to "people" dramatic tableaux and comment on social mores—a sort of "ani-drag." This near-total abnegation of the nonhuman image, Berger argues, served as a precursor to "the banality of Disney."[12] In Grandville, he contends:

> animals are not being "borrowed" to explain people, nothing is unmasked; on the contrary. These animals have become prisoners of a human/social situation into which they have been pressganged. The vulture as landlord is more dreadfully rapacious than he is as a bird. The crocodiles at dinner are greedier at the table than they are in the river. … The dogs in Grandville's engraving of a dog-pound are in no way canine; they have dog faces, but what they are suffering is imprisonment *like men* (1991: 19, emphasis in original).[13]

Ani-drag has predominated in the medium, across genres, since the modern form's emergence in the late nineteenth century; as noted by Michael Chaney, "comics have theorized the animal by *performing* it since their incipience and contemporary productions show no signs of flagging in this regard."[14] Whether in such classic figures as Otto Messmer's Felix the Cat, Carl Barks' Donald Duck and family, or "edgier" and more recent anti-heroes such as Robert Crumb's Fritz the Cat, animals in graphic narrative often partake of what Lisa Brown terms stereotypical "reverse anthropomorphism," its features codified by artists such as Will Eisner (inspired in part by the seventeenth-century physiognomic studies of Charles le Brun). She writes: "[I]mages of animals are culturally coded and take on certain prescribed characteristics – especially in comics: the proud lion, the mischievous cat, the sly fox, the wise owl, the dumb bear, the untrustworthy snake."[15] Or as Chaney concludes: "The animal in such comics always functions as mask or costume, beneath which lies the human, whose universality is affirmed and reified in the process."[16]

Animals' hyper-(in)visibility in such a visual economy (reflecting their utter marginality) arises from their perceived radical otherness and concomitant speechlessness—as Jacques Derrida famously put it, their "bottomless … uninterpretable, unreadable, undecidable, abyssal and secret" gaze.[17] Such a perception extends beyond the Western developed world, as evinced by Rupert Bazambanza's 2005 graphic novel *Smile Through the Tears,* based on the author's experiences in the 1994 Rwanda genocide. In his discussion of Bazambanza's work, Chaney registers that animals merely bear witness to the extreme limits of (human) experience; they appear as a "recurring evocative image of traumatic narration."[18]

Even the recent wave of animal-related or animal-centered comics works discussed by Chaney, David Herman and others—including writer Grant

Morrison's 1980s run on the DC series *Animal Man;* Stephen Murphy and Michael Zulli's *The Puma Blues* (1986/2015); David B.'s *Epileptic* (2000); Morrison and Frank Quitely's *We*³ (2005); James Vining's *First in Space* (2007) and Richard McGuire's *Here* (1989/2015)—while all clearly of a different order in terms of their environmental consciousness (often in a solemn, eco-catastrophic vein),[19] must still grapple with Derrida's dilemma of on some level "speaking for" the "dead" animal or "animort."[20]As Jonathan Burt has written in regards to cinema: "The visual animal is caught in an argument over whether the animal should be considered on its own terms or understood through a network of human-animal relations"[21]—an argument animals almost inevitably lose.

A voiceless, radically marginalized figure for whom anyone may speak recalls another such object of longstanding affective and ideological potency in Western culture: the corpse.[22] As anthropologist Katherine Verdery argues, "what gives a dead body symbolic effectiveness in politics is precisely its ambiguity, its capacity to evoke a variety of understandings."[23] Indeed, for most of the medium's history animals have suffered repeated deaths through the (mis-) representational ventriloquism of ani-drag (animals made to signify whatever the author chooses), reflecting their "dead" peripheral status in the larger world.

The foregoing underscores what Sarah Bezan identifies as a prevalent "framework of finitude" in Critical Animal Studies, present too in animal-based comics for much of the medium's history (if only in latent form). Bezan sees in "finitude" a morbid if politically useful tool for advancing "a responsible and compassionate ethical paradigm that responds to the biopolitical (or thanatopolitical) state's violent and systematic slaughter of human, and especially nonhuman, animals."[24]

In sum, this brief excursus has sought to show the animal figure in graphic narrative as fully subsumed into what Chaney calls a "strategic and parodic veiling of the human," a nonliving façade of otherness above which there often looms "the shadow of death."[25] In this regard, the animal *qua* animal disavowed by representation becomes available for mourning—at least for those willing to resist representational convention, to trouble the textures of said façade. Abadzis' *Laika*—a work which devotes page after page, with unprecedented naturalistic detail, to the physical/spiritual travails, abuse and destruction of an animal—is just such a resistance, just such a troubling.

LAIKA

Abadzis' graphic novel, an East-West parable of fraught human relationships with the animals they both love and cruelly exploit, generally adheres to the historiographic record, with some imaginative sequences to enhance drama and ground the reader in nonhuman species subjectivity. Kudryavka ("Little Curly," due to her tail) is born sometime in 1954 to a relatively affluent household, whose mistress commands the entire litter be gotten rid of (*"Drown* them if you have to!").[26] The solicitous mutt is briefly adopted, then passed on by the household servants; suffers abuse from an adolescent who tries to drown her in the Moscow river; lives on the streets with a canine companion, Gertruda (later beaten to death by a city dog catcher); and in dreams of flying[27] reunites with her original owner, the girl Liliana, with whom she seems to share a near-telepathic link. (This entire speculative "prequel" takes place in Chapter 2, amounting to some 40 pages.)

Eventually caught by dog catchers, Kudryavka winds up in the Institute of Aviation Medicine in 1956, where we reenter the realm of history proper. We follow her and other canine cosmonauts' grueling regimen as they train for their missions. The reader already knows what they are preparing for: in Chapter 1, father of the Soviet space program Chief Designer Sergei Korolev (whose 1938 release from a Soviet labor camp opens the novel and who to some extent parallels Kudryavka's own narrative journey)[28] has been ordered by Premiere Nikita Khrushchev himself to follow up his October, 1957 launch of the first satellite, *Sputnik,* with another socialist triumph: to put a living being in orbit in time for the 40th anniversary of the Bolshevik Revolution. The ludicrous one-month deadline means the space capsule's passenger will not return.

It falls to the imperious Korolev to choose which dog will make the one-way journey. He selects Kudryavka, bestowing a new name, Laika, on the mutt due to her barking (from the verb *laiat',* to bark).[29] Besides Korolev, Abadzis includes several other real-life figures, such as the aforementioned Gazenko. But while many scenes in *Laika* have a basis in actual people and events—as when Dr. Vladimir Yazdovsky takes Laika home to play with his kids shortly before her flight (discussed further on)[30]—Abadzis invents one crucial major character, the institute's dog handler Yelena Dubrovsky. Her deep bond with Laika and their heart-wrenching separation make for the book's most poignant and affecting scenes.

All in all, the author's orchestration of fact and fiction succeeds brilliantly as a plausible version of history and as a tragically cathartic tale of an animal's gradual annihilation punctuated by moments of profound tenderness, focalized through human and nonhuman perspectives. As one of the graphic novel's more perceptive critics, Rich Kreiner, writes, "Abadzis relates these annals with the pessimism they deserve,"[31] never denying the ultimately morbid and doleful nature of his material. This makes only more crucial the role played by invented foreshadowing incidents "to broaden and deepen the reader's emotional connection." Kreiner goes on: "The thematic links are critical in generating a more profound sympathy for what essentially is—however otherwise dependent, anxious to please and lovable—a canine cipher."[32] In other words, for all the compassion a reader may feel for Laika, the endearing, compliant mutt—and at times she indeed seems dangerously close to saint-like—Abadzis insistently presents the animal as a material body, a nonhuman creature of flesh, blood, and vomit, a living being which deserves consideration precisely as an irreducible Other. A "cipher" in Kreiner's terms; the "irreducible living multiplicity" of the *animot* in Derrida's.[33]

Abadzis pursues this task through sustained visual-verbal strategies to de-sentimentalize and de-anthropomorphize his subject.[34] As he said to an interviewer:

> Originally, very early on in the process ... Laika talked. That very swiftly got thrown out; I don't think I was ever actually serious about that. Empathy and sympathy and working to evoke those from the reader are good; sentiment and anything twee isn't. Laika is already cute and she's a dog with dog behavior; it was tough but one of the things I was sure about was if she was going to be anthropomorphized then it should be, as far as possible, through the eyes of a human character.[35]

The artist's stated approach has striking consequences for the novel's formal construction, as well as for the reader's involvement. *Laika's* shifting, often ambiguous, point of view on who is speaking, dreaming, and experiencing draws on the fundamental open-endedness of the comics gutter (the "empty" space between panels), which each reader here imbues with her own content, to motivate the narrative and blur lines between selves and species.[36] In the early dream sequence, for example, we see Kudryavka settle down to sleep, followed by a series of wavy-bordered panels against solid black, which show her consorting with her former owner's mother Tatyana and flying through the starry night.[37] Over the next two pages she flies on, encountering figures from her life until, against a white background, the panels take her directly into the arms of Liliana, her beloved first owner. On the final panel of page 53, they are joyously reunited against a field of yellow. We can take that color as a marker of waking—for on the very next page we discover Liliana in bed, roused from sleep. She goes on to tell her mother the exact dream we have been following, leading readers to question who in fact was dreaming all along: girl, dog, both?[38]

In addition to such panel-sequence, page-design, and page-turn tactics, Abadzis destabilizes voice and subjectivity by capitalizing on comics' fundamentally *ironic* pairing of word and image, what Hillary Chute has called its cross-discursive poetics and Charles Hatfield an aspect of its "art of tensions."[39] Thus, what seems fairly straightforward becomes increasingly obscure and cryptic—such as who is saying what. In an early scene, after losing their dog forever, Liliana's mother tries to calm her worried daughter by asking, "If she could talk, what would you ask her?," to which the girl replies, "Are you all right, Kudryavka?" Tatyana responds, her dialogue in quotes, "I'm fine, Liliana." The next panel shows an imagined Kudryavka, her head floating over the girl, continuing to console her. Yet though the speech balloon's tail points to the mutt, her dialogue retains the quotation marks. As Liliana falls asleep, comforted, we see her mother in the next panel, continuing to ventriloquize Kudryavka's speech, still in quotes.[40] The scene powerfully advances one of the novel's chief themes: the human characters' presumption, habit—compulsion, even—*to speak for* animals.

Abadzis' most developed invented heroine, Yelena Dubrovsky, elevates this tendency to a troubling, quasipsychotic degree. Hired by the institute due to her easy facility and rapport with the animals, Dubrovsky breezily declares on her first day, "Yes, dogs *talk* to me!"[41]We soon learn she at times believes this literally, calling into question both her mental health and her ability to see the animals as they are. Staring at one of her charges, Albina, through the bars of the dog's cage, Dubrovsky asks, "What would you say if you could talk back?" In the next panel, startlingly, Albina seems to answer, "Let me out." The following frame, the third on the page, registers Dubrovsky's surprise through its truncated height, which leaves a relatively wide gutter with the tier/panel beneath. She really thinks she is having a conversation with the dog. But again, as in the previous scene, the dog's "dialogue" appears in quotes, signaling its imagined, projective status (to the reader if not to Dubrovsky). Another clue: Albina's speech balloons lack a tail; they are unattributed and thus open to interpretation.[42] A later, even more disconcerting scene depicts all the dogs clamoring in their cages for release from their prison, each begging Dubrovsky to spare her from her fate and take her home. At first "speaking" through tailed speech balloons with quoted text,[43] by the next page's first panel we see a jumble of untailed speech balloons, each in its own color, with no quotation marks encasing the words: "Me!" "Me!" ... "Please, take us home!" ... "Please – me!" ... "Me!"[44] The lab's lighting too, has altered, from after-hours dark to a brightly lit, "yellow-wallpaperish" hue of teetering sanity. By close to the novel's end, Dubrovsky (fueled by alcohol) no longer resists her visions; she has extended "conversations" with Kudryavka and Albina at the site of the satellite launch (in remote Kazakhstan). Kudryavka/Laika calls the place "a freezing wasteland," telling Dubrovsky, "It's not fit for a dog, much less for *you*, Mistress Yelena!"[45] Once more, the animals' dialogue appears attributed—but without quotes. Do they, then, speak? Should we continue to read these as Dubrovsky's ravings, or as some genuine communication across species? Or some odd combination of both?

Abadzis, through the willful imprecision of "no one's" speech balloons, leaves the question hanging; the dialogue does not necessarily emanate from the animals, without altogether foreclosing such a possibility. These scenes partake of what comics theorist Scott McCloud calls the medium's interdependent visual-verbal pairing, in which "words and pictures go *hand in hand* to convey an idea that neither could convey *alone,*" even when the words seem to say one thing and the image another. The author uses that contrapuntal quality of graphic narrative to forcefully remind the reader of the subject status of animals, who as J. M. Coetzee writes, "have only their silence left with which to confront us."[46] A silence whose vacuum people fill with alacrity—perhaps unavoidably. Abadzis here expresses through comic art a cautionary ethics of encounter with the animal figure, an ethics which seeks to overcome what David Herman calls, "the anthropomorphic bias of narratives" and constitute story-worlds "on different grounds"[47] through double-voicedness, confusion, outright inapprehension.[48]

More conventionally, Abadzis generates sympathy for his animal protagonist through the scientists' (mostly Dubrovsky's) excruciating expressions of grief as Laika trains, gets locked up in the spacecraft, languishes in the overheated capsule and ultimately dies. *Laika,* a story "steeped in fatalism" yet commanding "authority as a historical chronicle,"[49] thus confronts readers with another major theme, summed up by reviewer Lisa Brown: "To what extent, if any, does human ambition outweigh a life?"[50]

Such a "framework of finitude" approach links the novel to decades of mournful, morally fraught Laika discourses which as noted have pervaded the subject since the beginning. Nelson, for example, describes a Moscow planetarium audience's "collective sigh" when news of the dog's death was announced in 1957.[51] But, in a reminder of Butler's "oddly fecund" power of loss and pathos, such discourses come to function ahistorically, retroactively as anticipatory grief in representations of Laika that show her before her death.

We see a stark visualization of such collective anticipatory grief in the novel when Dubrovsky first learns of Laika's doomed flight. Page 143 shows Yazdovsky revealing the truth to his rueful colleague: in the third frame, he tells her, "She won't be coming back. It's unfortunate. I'm sorry." Abadzis isolates Dubrovsky's figure by placing it, waist-up, at the extreme right of the page-wide rectangular panel, further set off by her white lab coat (contrasted to Yazdovsky's dark suit); by a transverse shadow falling on the extreme left, leaving her illuminated and exposed; and, most strikingly, by her face, which suddenly appears blank, completely whited out. In the next two panels, as Yazdovsky leaves, Dubrovsky's body undergoes even more "emptying": it appears first as a black then a white silhouette. The art's marked turn to expressionism signals the vacuity of grief, how it transcends physical/temporal barriers, seeping everywhere to sap the very surroundings of life, character, detail—even one's body comes to feel alien, hollow, a mask.

But this only accounts for part of the page. In the remaining four panels, all wordless, all descending in a column on the right, we see Yazdovsky at home, watching his children play with Kudryavka (he wanted to give her a taste of life as an ordinary dog while he could) and, in the final panel, Gazenko at his usual watering hole, also grappling with his feelings about the imminent mission. Through comics' juxtapositional strategies—arranging disparate time periods, locations, moods, focalizations side by side—the scene communicates how different people (the family man Yazdovsky, the stoic bachelor Gazenko, the devoted dog-handler Dubrovsky), each mourns in their own way.

As hinted here, and as borne out repeatedly in the novel, Abadzis limns reactions to loss (anticipated and actual), along with animal-human bonding, in notably gendered terms. A repeated motif shows Kudryavka in the arms of her original owner Tatyana, mirrored in Dubrovsky's own embrace of the dog.[52] The portraits recall a particularly Russian Madonna and child (Tatyana typically wears a babushka), not unlike the painter Kuzma Petrov-Vodkin's *Petrograd in 1918* (aka *Petrograd Madonna,* 1920). Such imagery strongly clashes with the barrel-chested masculine swagger of Korolev and Khrushchev; the maudlin "beta male" passivity of Gazenko; and the soulless bureaucracy of the Soviet state, symbolized by the hyper-phallic, burning red rocket primed to propel Kudryavka to the stars, which Abadzis renders in its own splash panel.[53] In *Laika's* gender economy, women consistently feel, emote and suffer loss better than men. Dubrovsky suffers most, falling into a drunken, hallucinatory depression as Laika's launch approaches.

And indeed, the novel's most unsettling deviation from conventional Laika discourses since 1957 (heroic, sad, clean) unfolds in Abadzis' depiction of the harrowing flight itself. As the rocket leaves Earth, the latent is rendered brutally explicit: we bear witness to the dog's physiological torments, her moment-by-moment agony aboard a "scorching coffin."[54]

David Herman, who has examined this scene at length, proceeds from the work of the philosopher-biologist Jakob von Uexküll to emphasize its layering of four different "experiential frames," each conveying a different subjectivity or focalization: the "dominant frame" of the rocket in space; Laika's moment-by-moment telemetry readings; Dubrovsky and the other humans' view; and the dreams/fantasies/memories/stream of consciousness of Laika and possibly Dubrovsky.[55] Herman relies on von Uexküll's concept of *umwelt*—the lived, perception-saturated phenomenal world of nonhuman living things—to argue for the efficacy of narratives (including graphic narratives) which resist anthropocentric biases to convey a sense of that alterity. "By modeling the richness and complexity of 'what it is like' for nonhuman others, stories can underscore what is at stake in the trivialization – or outright destruction – of their experiences," Herman writes, with *Laika's Sputnik-2* flight as his model.[56]

Furthermore, Abadzis' depiction of Laika's subjectivity in the sequence lies at a double-limit of representation: the umwelt of the animal and the experience of dying. To the extent that humans have no access to the former and only those actively in the process of dying themselves can access the latter (albeit through their own subjectivity; we can each only experience our own death), this representation admits more than a little of the colonialist—if not, again, of the outright destructive. As Elisabeth Bronfen contends,

> [Is] every representation of dying not violent precisely because it implies the safe position of a spectator ("voyeur") and because a fragmentation and idolisation of the body—i.e. a severing of the body from its real materiality and its historical context ("fetishism")—is always built into such images. The body—always already the site of exterior and alterior inscriptions – is explicitly exposed as such an inscribed site when it is in the process of dying, so that in representations of the dying body this "alterior" inscription is doubled through aestheticization.[57]

Or, arguably, tripled in the case of a dying animal. In her discussion, Bronfen references the Swiss painter Ferdinand Hodler, who from February, 1914 to January, 1915 produced over 70 portraits of his lover Valentine Godé-Darel as she was wasting away on her deathbed. For Bronfen the highly "aestheticized" sketches and paintings come to stand in more for the artist's own processes of grief than for any "fidelity" to their subject—the proof of this comes six months after Godé-Darel's death, when Hodler paints a new portrait of her alive and healthy. The beautiful remembrance trumps the "real" subject of the dying body—Bronfen's "exterior and alterior inscriptions" have done their work. One "los[es] the real so to speak to the needs of a representational unity," she maintains.[58]

As noted, something akin to this unavoidable aesthetic colonization happens in *Laika* as well—but I submit that Abadzis, through comics' textual-pictorial and juxtapositional-sequential schema for meaning-making, renders his version of the canine cosmonaut's death more complicated, ultimately more self-reflexive and ethically defensible than in the Godé-Darel case. For one thing, comics have words which interact with, challenge, and again ironicize the images. Repeated throughout the novel, the expressions, "You can trust me," "Good dog," and "Don't worry" reappear in the death scene, in panels that show the beloved mutt suffering unconscionably.

Damning and hollow, these recurrent phrases—spoken by various characters in the course of the story—mark the banal betrayal of animals by people. But their appearance (unattributed) during the mission reinforces all the more the mood of atrocity and sadism as a terrified animal is cooked alive. Who speaks these words? Does Dubrovsky telepathically transmit them? Does Laika recall them from memory? Does some other narrative voice (the reader's?) intercede in a vain attempt to comfort the dog? Or may we simply read these platitudes as a marker of the "canine cipher's" loyalty onto death? Or all of the above, and more: these hackneyed utterances gesture at another betrayal—the one effected whenever we seek to represent the dying, the animal, or in this case a dying animal. Graphic narrative's word-image tension allows Abadzis to comment on his own "killing" even as he carries it out.

Moreover, the unfurling of Laika's execution over several pages, in a facsimile of real time due to its embedding in an established narrative with multiple perspectives and the comics panel's spatio-temporal ambivalence means we cannot reduce the dog's body to any one meaning or "alterior inscription." As Groensteen argues, "Each new panel hastens the story and, simultaneously, holds it back. The frame is the agent of this double maneuver of progression/retention."[59]

The page on which Laika/Kudryavka dies (though intriguingly, none of the scientists monitoring the capsule actually declare her dead) shows two narrow horizontal panels extending the full length of the page, bleeding to the material ends of the paper. These depict the dog's tiny black silhouette against a white void (panel one), and the white void sans dog (panel two). We may read these polysemantic panels in all sorts of ways and tie them to many different subjectivities (or none), but the meaning remains basically clear: our heroine is no more. The next tier holds one square panel of total black with a green border (green bordered panels up to now have served as Laika's "experiential frame"), floating in gray empty space. Immediately below the black panel: another rectangular vertical frame, nearly half the length of the page, which also bleeds to the bottom of the book border. This panel shows a devastated Dubrovsky in tears, her arms achingly empty, a crushing reminder of the absent dog. She stands helplessly against a field of deep lavender,[60] above which floats an inky blackness mirroring that of the black panel. The visual intersection/linkage of Dubrovsky and Laika at the "moment" of the latter's end again suggests their near-supernatural connection, but it also simply depicts the sundering of ties brought about by the loss of a loved one.[61]

In short, even if Laika, especially in death, falls victim to the "alterior inscriptions" of others—rendering her a poster dog for the USSR's inhumanity, say, or a heroic "sacrifice to science," or a stand-in for man in an increasingly mechanistic modernity—in comics form her representation, with its numerous "experiential frames" and cross-discursive tensions, does not reduce her to mere symbol in any such instrumental fashion (too much else is going on, pictorially and narratively). Instead, the reader witnesses her death first and foremost as barbaric, unjust, repulsive—the death of an actual, innocent being. Even Kudryavka's final fate at the end of the graphic novel—she becomes falling stars—resists any "symbolic" transfiguration (we see no parent telling a child to "make a wish"). The oddly precise date

("April 13,1958," in a text box in the first panel); the "scientific" sequence of frames showing what happens when a spacecraft's orbit decays and it reenters the Earth's atmosphere; the lack of commentary, all militate against any one reading.[62]

That said, after the comics portion Abadzis inserts Gazenko's rueful 1998 disavowal of the mission, followed by a photo of the historical Laika (as part of an afterword). The black and white photograph in particular throws the reader back to a confrontation with the flesh-and-blood figure of which, we are reminded, we have only been engaging with on an imaginary, speculative, ink-and-paper level. Kreiner calls the inclusion of Laika's photo "savagely eloquent," for it "effectively supplants her drawn representation in suggesting the weight and woe of her story."[63]

Such a return of the "real" at book's end, perhaps in some measure, contributed to the graphic novel's odd sequel.

"AFTERLIVES"

In July, 2011, Abadzis posted to his blog:

> It's come to my attention over the years since my book *Laika* was published by First Second in 2007 that the ending upset a lot of readers.
>
> The truth is, even if they've never heard of this little book, anyone who's heard of Laika has googled her and found out how her story ends. But somehow, we wish we could change it. We wish we could change the way things happened by writing about them.
>
> Many people still contact me, wondering about that ending, wondering why I told the story the way I did, about how much of it was real and even whether there might be an alternative ending to the story … . Filmmakers often get in touch, wondering whether there might be a way of presenting a version with a more positive spin to it. Well, of course there is, but then you'd be changing history, or at least blunting the truth of what took place that day in 1957. And unfortunately, you can't change history, not one line.[64]

The uproar had precedents in comic history. Mary Gold, an ongoing character in Sidney Smith's enormously popular early twentieth-century newspaper strip *The Gumps,* died after a long illness in the episode published on April 30, 1929. This first cartoon death, according to Jared Gardner, provoked an "avalanche of mail"; fans pleaded with Smith to change the outcome and in some cases repudiated him for his refusal: "[P]apers reported the fervent hopes of their most desperate readers that Mary might not actually be dead. One speculated that it was only a dream. Another that she had merely, "fallen into a natural sleep."[65]

A more recent and relevant case of "cartoon death" unfolded in the funny papers in 1995, when Lynn Johnston, creator of the daily syndicated comic strip *For Better or For Worse,* decided to kill off the beloved Patterson family dog, Farley, in a heroic act of sacrifice. (The old English Sheep Dog rescues a family member from a raging river, then dies of a heart attack.)[66]At the time, after some 15 years of publication, Johnston had achieved tremendous success and earned a loyal fan base.

These fans exploded in grief and outrage, writing in extraordinary numbers to register their protest or express profound sadness at the turn of events. "People's emotions were kind of raw," said Johnston at the time. "I received 2,500 letters, about one-third negative. I didn't expect the response to be so great. The letters were open and emotional and honest and personal, full of stories and love."[67]

In one of these letters, the reader demanded Johnston set right what they perceived as a cosmic wrong:

> We adore the Patterson family—the whole family—that includes Farley. We're certain they are living and breathing up there somewhere in Canada. Living and BREATHING! That is exactly what we want Farley to go on doing. Don't you dare kill him. Let him get very ill, have a near death experience, see white light if you wish, but, we implore you, Farley must not die. IF you have already drawn Farley dead, just rip it up and FAX a recovering Farley. Since the Pattersons are real, unlike the ageless Blondie and Dagwood and Daisy, it is only logical that Farley wont [*sic*] live forever. Just make an exception. Who cares about logic? Life is tough enough. Please don't add to our miserys [*sic*].[68]

The writer's presumptuous, pushy tone only accents all the more the tone of desperation. A number of the letters frankly discussed the personal loss of a pet: "[I] find the subject appalling, particularly since I'm dealing with the imminent demise of a 15 year old Labrador. Comics are for escapism, Ms Johnston. They're not supposed to be a slap in the face with reality."

In part reacting to such outpourings of grief, in April, 1995, Johnston produced a long-form Sunday strip in which Farley is commemorated—and visually idealized. Strolling in a wooded area, Elizabeth Patterson laments to a friend, "I didn't think I'd miss him like this, Dawn. I never thought that a dog could mean so much." They vow to remember a treasured spot as "Farley's tree." In the concluding panel, we see them followed by their shaggy friend, now in "ghostly" form. "Sometimes I hear the sound of his collar … . As if he was walking right here," Liz says.[69] The artist's integrity towards her subject here finds a compromise with the fan imperative to "make things right," through fantasy.

Abadzis, in his own response to fans' demands to "make things right," even if "you can't change history, not one line," chose a less maudlin and more self-aware strategy, though one too rooted in fantasy. In August, 2011, he posted online four alternate endings to *Laika,*[70] all giving his readers what they wanted: their beloved mutt alive and well. These extraordinary "addenda" range from the almost sort-of plausible (the scientists arrange a parachute escape; the dog joyously licks Dubrovsky's face in reunion) to the morally suspect (at the last moment, another canine cosmonaut, Albina, is substituted for Kudryavka; the former dies so the latter may live, though she proves so inspirational that the Soviets beat the United States to the moon), from the mockingly referential (in space, Kudryavka is rescued by an alien "monolith" which transforms her into a "star puppy"—an allusion to the Arthur C. Clarke classic sci-fi novel *2001: A Space Odyssey* and the Stanley Kubrick film adaptation, both 1968) to sardonically fannish (the intrepid canine not only survives in orbit but, through exposure to "cosmic rays," becomes a caped superhero, Cosmodog, hungry for vengeance against his Soviet tormenters ("They fear *power,* they fear *me.*"))—a parody of both the Stan Lee/Jack Kirby Silver Age series *Fantastic Four* and the wooden dialogue of much contemporary superhero comics writing.[71]

Clearly, disavowal of life's harsh realities and wish fulfillment play major roles in these supremely ironic, hilariously fake alternate endings. (The convenience of the internet as a medium for distribution also seems a factor in their click-bait, "you can't be serious" disposability.) And, in fact, they had their origins not only in fan letters, but in the Dream Factory itself. As Abadzis told me:

> The alternative endings to *Laika* came about when there were a few Hollywood agents and questions about movie rights around. I kept being asked if I could change the ending. I said that I could put a spin on it—a coda, perhaps, to demonstrate the fact that the Russians sent up further cosmodogs, whom they took great pains to get back, which was in part a reaction to the PR problem that the Laika mission represented. One of the successor cosmodogs had puppies and one of these was given by Khrushchev to President Kennedy, so it ran about on the White House lawn for a few years. I thought that, if you wanted to put a cute Hollywood ending on the story of Laika that would be a way to do it. Those experiences were the genesis of the idea for the alternative endings.[72]

Functioning much more readily as parodies, however, the new endings ultimately puncture the crass will to fantasy of the "Hollywood happy ending" rather than uphold it. Further, one detects here a whiff of satire not only of death-denying generic conventions in mainstream films and comics, but perhaps of readers' own demands to "make it right," to refuse to face up to the truths of history—such as Laika's horrific fate, or the mass abuse of animals, much less to one's own mortality. In short, these new endings subtly repudiate the puerile impulse for easy escapism from mourning. When I suggested as much to Abadzis, he demurred: "Did some humor creep in to those alternative endings? Sure. I'm a cartoonist, and despite the overall tone of the book, I'm capable of different moods."[73]

Shortly after posting the four alternate endings, in 2011 Abadzis produced what seems his final word on the subject: a comics essay, an epilogue of sorts to the novel, in which he addresses his own mixed emotions.[74] This "fifth ending's" art ranges from the photorealistic (Kudryavka is drawn in greater detail, from photographs) to hyper-cartoony (learning about the doomed space mutt, a child Abadzis pictures her as Charles Schulz's Snoopy, prone atop a red rocket, his eyes x'ed out), mirroring the flux of feelings provoked by the Laika case. In the strip, the author eulogizes the dog's flight and acknowledges the enormous pressures put on Soviet scientists to get results. He also honors the need for death-denying fantasy, admitting that when younger he imagined Laika rescued by aliens and living out wonderful exploits on the planet Fabuluxxus.[75] Figuring himself at his computer and drawing table, he muses, "All those notions are fun and ease the knowledge of her true fate. Thing is, why should anyone worry about the death of a dog from more than half a century ago? There are animals, and people all over the planet treated inhumanely every single day, as they have been since Laika flew."

In this "framework of finitude" vein, Abadzis concludes by contemplating a famous photograph of Laika in her space suit (the very one which illustrates the graphic novel's afterword), before walking through a door with the dog by his side (she shifts in her framed photograph and seemingly leaps out of it between panels). The door leads to outer space, a comet blazing in the distance. These last two panels read: "We all stand on the shoulders of those who made sacrifices, or were sacrificed whether they agreed to it or not. It's the reason the luckiest of us live freely. Is there an alternative ending to that?"[76]

Abadzis' glum closing appeal to the "we," to the "us" (homo sapiens?) on the supposed benefits derived from the unlucky "sacrificed" few (presumably animals of all sorts, including human beings) seems to both reinforce species barriers and dissolve them through empathy. It returns "us" to Judith Butler's writings on precarity and grievable lives, the blurring of identities catalyzed by bereavement, which Butler sees as a recognition of the Other's profound imbrication with the self: "Perhaps, rather, one mourns when one accepts that by the loss one undergoes one will be changed, possibly forever. Perhaps mourning has to do with agreeing to undergo a transformation (perhaps one should say *submitting* to a transformation) the full result of which one cannot know in advance."[77]

Or as Stanescu stresses, mourning both generates and casts into relief the variegated kinship ties that shape individual selves and communities, crucially premised on "the vulnerability and finitude of the other,"[78] which is also "our" vulnerability and finitude. He goes on: "The beings whose lives are gone still continue with us when we mourn them. Their reality continues to impose itself on our reality; their existence continues on in our existence."[79]

Abadzis' work, as attested to by the plaintive protests it inspired, taps into the potency of such affectual afterlives in the Laika case itself, and of graphic narrative's capacities to represent it. In her discussion of the novel's alternate endings, blogger Becky Ferreira writes,

> The fact that Laika can still inspire this palpable desire for revisionist wish-fulfillment 57 years after her death speaks volumes about her continued relevance as an ideological lightning rod. Dogs are the undisputed champions of loyalty and love, and it will always be painful to accept that Laika's blind trust in her masters was betrayed.[80]

Beyond the mawkish anthropocentrism, this commentary rightly points out the centrality of "making it right" in Laika discourses[81] of the past, present, and presumably future. And a future they definitely have. As Turkina *et al.* opine: "Gradually, the perceptions of the Cold War and the opposition between socialist and capitalist systems will begin to fade from our memory; eventually the acronym 'USSR' will become a mystery; yet most people on Earth will still be able to recall Laika's name and image."[82]

But what will that name and image mean?

CONCLUSION: WITHOUT WORDS

As for animals being too dumb to speak for themselves, consider the following sequence of events. When Albert Camus was a young boy in Algeria, his grandmother told him to bring her one of the hens from their cage in the backyard. He obeyed, then watched her cut off its head with a kitchen knife, catching its blood in a bowl so that the floor would not be dirtied.

The death-cry of that hen imprinted itself on the boy's memory so hauntingly that in 1958 he wrote an impassioned attack on the guillotine. As a result, in part, of that polemic, capital punishment was abolished in France. Who is to say, then, that the hen did not speak?[83]

"Laika's tragic death, her self-sacrifice [*sic*] in the name of the new space religion (even though no one had asked her consent), and the very image of a defenseless little dog courageously zooming into the unknown, all contributed to her becoming the first icon of space exploration," note Turkina *et al.* "The dog's valiant little face with its pointy ears appeared on the front page of every newspaper around the world. Her poignant image came to symbolize both heroism and the undying human capacity for compassion."[84]

Few would dispute such a characterization, though a primary feature of what I have called Laika discourses is their heterogeneity, as I tried to make clear through a reading of Abadzis' compelling, formally sophisticated graphic novel. Yes, Laika means that. And that. And this.

Such "multi-voicedness" on the one hand comprehends the stochastic and manifold nature of mourning; as Helen MacDonald eloquently puts it: "The archaeology of grief is not ordered. It is more like earth under a spade, turning up things you had forgotten. Surprising things come to light: not simply memories, but states of mind, emotions, older ways of seeing the world."[85] On the other hand, the proliferation of "heroic," "tragic," "sacrificial" speech about this one animal's short life and cruel death only underscores the fact that she herself had no voice.

To be blunt, everything said and done to commemorate her—from the first memorial, a brass tag placed on her kennel[86] to Abadzis' solemn fifth ending—kills her anew, points to her fundamental "non-power," in Derrida's phrase. Moreover, the myriad "clean" and "kid-friendly" visualizations of Laika over the decades, precisely by smoothing over the gross and graphic details of her tortures, make her all the more available for consumption. Chloe Taylor says it more archly: "I would suggest we censor the sight and sounds of animal deaths because we need to keep animal lives and deaths derealized in order to continue with our plans."[87]

Even as a "good," politically useful symbol (of injustice towards animals; of their abuse; of the need for rescue; of man's disengagement from nature in an age of ecological collapse, etc.), Laika as Laika is denied, forgotten, unknown, a cipher. Loss remains at the center, a void as chill as space; you can love Laika to death all over again. As I argued, works like Abadzis' novel (and addenda), which don't deny their own implication in these exploitative processes but self-reflexively embrace them, offer readers perhaps the best case for a version of engagement.

Graphic narrative, with its inborn cross-discursivity (a two-headed chimera) offers even more such opportunities to get at the multivalent meanings and hopelessly mixed emotions we moderns feel towards those others with whom we share the planet. Comics, I submit, rivals any other medium in taking on the task of reading, what Colleen Boggs describes as,

> an act of encountering the bodies of others and of needing to come to terms with their proximity and alterity. This encounter with animal bodies challenges us to expand our understanding of the work sentimentalism can do for cross-species relations. ... What happens when sentimentalism does not automatically link back to humanity but instead creates connections with animals that press us beyond the human and humanist pale?[88]

In conclusion: one more of Abadzis' (missed) connections with an animal that presses beyond the humanist pale. Shortly before murdering her, Korolev subjects Kudryavka[89] to a monologue. "I'm sorry you're in that cage instead of running free," he begins. "I was a prisoner once." The Chief Designer tells of his arrest, internment in the GULAG system, and other Soviet horrors.

These recountings unfold over three and a half pages, in thin panels (mostly horizontal) the width or height of whatever page they inhabit. Korolev and Laika, in all these 21 panels, remain at opposite extremes, seemingly as equals. Yet over the course of the stilted one-way conversation, the dog is steadily abstracted, her figure visually yielding to the man's (mis)perceptions and memories. When Korolev relates how "the barking of a dog" helped him find his way to civilization from the northern wastes after his release, Laika is (impossibly) shadowed as if she were with him then, the moon from that night shining on her face.[90]

The animal's de-realization continues on the next page: her image is voided, emptied out, a "blank" white silhouette upon which Korolev projects his own fancies. "*You* spoke to *me,*" he says.[91] He wraps up his oration thus, speaking once more to Laika as an ordinary dog:

> So, for what I must do, I *apologize.* I *must* do this. If I do succeed, then maybe I will be free.
> And you will be, too.
> You will fly further and longer and higher than any living being from this Earth ever has.
> *I* am a man of destiny.
> And *you* will be the most famous *dog* in history.[92]

With that, the strutting Korolev leaves the lab. We are left with a small, rectangular panel (very different from Korolev's absurdly phallicized frames) of Laika in her cage. All this while she has stood quietly as the man droned on.

> Now she stares, from behind the bars, after Korolev, but also (a little) at the reader.
> Silent, with an inscrutable expression.
> Whatever she's saying—if anything—she says it without words.

For Gesolmina, Eleanor Roosevelt and Mabel, our first-hatched

NOTES

1. See Amy Nelson,"The Legacy of Laika: Celebrity, Sacrifice and the Soviet Space Dogs," in *Beastly Natures: Animals, Humans, and the Study of History*, ed. Dorothee Brantz (Charlottesville, VA: University of Virginia Press, 2010), 210, and Turkina, Olessya, Cannon, Inna, and Wasserman, Lisa, *Soviet Space Dogs* (London: FUEL Design & Publishing, 2014), 89.

2. Turkina *et al., Soviet Space Dogs*, 89.

3. Nelson, "Legacy of Laika," 215. The 2002 revelations inspired Nick Abadzis, author of *Laika,* to investigate further and eventually to create his graphic novel. As he explained in an interview: "It always struck me as odd and sad, that she'd been sent

up and they couldn't get her down again. Why? I always wanted to know the answer to that question," Coleen Mondor and Lisa Brown, "Animals in Space," *Antennae*. No. 16 (Spring, 2011): 11–12.

4. Turkina *et al.*, 131.

5. Nelson, "Legacy," 218.

6. Nelson reports some of these websites link Laika's story to the grief of pet owners (220); she also lists scores of musical groups (e.g., Laika and the Cosmonauts, from Finland) and pieces referencing the space mutt in "The Music of Memory and Forgetting: Global Echoes of Sputnik II," in *Remembering the Space Age*. NASA SP (Series) 4703. Ed. Steven J. Dick (Washington, DC: National Aeronautics and Space Administration, Office of External Relations, History Division, 2008), 239–248. Poetry includes Caitlin Doherty's "To Laika," part of her 2012 chapbook *Satellites*.

7. Turkina *et al.* echo this language when they write: "These dogs are the characters in a fairy tale that was created in the USSR: they are the martyrs and saints of communism." (9–10).

8. Abadzis (b. 1965) earned a devoted alternative comics readership with his series "Hugo Tate." (1988–1994) in the journal *Deadline*. His work has also appeared in *2000 A.D.* and the DC Comics imprint Vertigo. *Laika* was his first graphic novel for First Second Press, since its 2006 launch a critically acclaimed publisher of comics and children's books.

9. The novel won the Young Adult Library Services for Teens' Great Graphic Novel Prize and the Eisner Award for Best Reality-Based Work, among other accolades.

10. James Stanescu, "Species Trouble: Judith Butler, Mourning, and the Precarious Lives of Animals,"*Hypatia*27, No. 3 (2012): 579.

11. Rich Kreiner, "Laika," *The Comics Journal*. No. 292 (October, 2008): 193.

12. John Berger, *About Looking*, 1st Vintage International ed. (New York, NY: Vintage International, 1991), 19.

13. Ibid., 19.

14. Michael A. Chaney, "Animal Subjects of the Graphic Novel," *College Literature*38, No. 3 (2011): 129. My emphasis.

15. Brown, "An Introduction to the Illustrated Animal," *Antennae*. No. 16 (Spring, 2011): 4.

16. Chaney, "Animal Subjects," 135.

17. Jacques Derrida, *The Animal That Therefore I Am*, ed. Marie-Louise Mallet, trans. Davis Wills. (New York: Fordham University Press, 2008), 12.

18. Similarly, commenting on Art Spiegelman's celebrated Holocaust graphic memoir *Maus* (1986–1991), which employs a "hybrid bodies" trope of mouse heads to represent Jews, Katalin Orbán argues "the empty gaze—and the animal head that serves the same purpose in Spiegelman's work—evokes not only a *connection* but also a *disconnection*. Besides the connection of blindness and under-standing, it also evokes that dissociation of sight and comprehension that is central to trauma and a recurring element in the Holocaust survivors' accounts." (63).

19. Certainly many critics read them in this mode; for example, see the conclusion to Michael Chaney's forthcoming *Reading Lessons in Seeing*, devoted in part to animals in McGuire's experimental graphic novel *Here*.

20. Derrida, *The Animal That Therefore I Am*, 62.

21. Jonathan Burt, *Animals in Film* (London: Reaktion, 2002), 188. Elsewhere he adds on this vexed issue of representation: "It is usually argued that the exploitation of visual imagery for the purposes of animal rights is due to the fact that animals cannot speak up for themselves, so the message is in greater need of visual reinforcement than, presumably, for issues of human rights. In fact, the actual power of this imagery derives from a much longer term concern over public codes concerning what should and should not be seen." (168).

22. See my discussion, after Elisabeth Bronfen and Sarah Godwin, of death's resistance to representation leading to the corpse's polysemic potential in comics in *Death, Disability and the Superhero: The Silver Age and Beyond (*Jackson, MS: University Press of Mississippi, 2014), 162–163 and Chapter 9.

23. Katherine Verdery, *The Political Lives of Dead Bodies: Reburial and Post-socialist Change*. Annual W. Averell Harriman Lecture (New York, NY: Columbia University Press, 1999), 29.

24. Though she calls it into question as corrosive of agency and on other grounds; she advances a different approach, proceeding in part from vitalist philosophy. See especially 191–194.

25. Chaney, "Animal Subjects," 130.

26. Unless otherwise stated, emphases in comics works are in the original.

27. Throughout the novel, dreams play a critical role in cementing cross-species connections, albeit fancifully. These oneiric sequences (Laika floating joyfully over a moon-filled nighttime Moscow, with blurring of identities, telepathic overlaps, etc.) recall Mikhail Bulgakov's classic novel *The Master and Margarita* (1940/1967). In their discussion of the historical Laika, Turkina, *et al.* cite another Bulgakov novel, *The Heart of a Dog* (1925), large portions of which are narrated from the dog Sharik's point of view (131–132).

28. As Abadzis told an interviewer: "In a way the whole story of Laika is begun by the inhumanities that are done to Korolev; it's a cycle of abuse that continues throughout the book." In Tom Spurgeon, "A Short Interview with Nick Abadzis," *The Comics Reporter* (September 1, 2007). http://www.comicsreporter.com/index.php/resources/interviews/9884/.

29. She is also called Zhuchka ("Little Bug") when she arrives at the institute (68). These rechristenings reinforce the theme of the animal's utter "deadness" and manipulability from an anthropocentric perspective.

30. See Nelson, "Legacy of Laika," 214 and Turkina *et al.*, *Soviet Space Dogs*, 61, 66, and 118 for accounts of Yazdovsky and other Soviet scientists growing emotionally attached to their animals in the space program.

31. Rich Kreiner, "Laika," *The Comics Journal*. No. 292 (October, 2008): 190.

32. Ibid., 190.

33. Derrida, *Animal*, 41.

34. In this, as mentioned, Abadzis was going against the decades-long grain of "humanizing" space dogs, going back to their very introduction to the public, which included "head shots" in the press (see Nelson "Legacy of Laika": 215).

35. Spurgeon, "Short Interview."

36. On the gutter and its function through closure, see Scott McCloud, *Understanding Comics* (Northampton, MA: Kitchen Sink, 1993), Chapter 3.

37. Nick Abadzis, *Laika* 1st ed. (New York, NY: First Second, 2007), 51.

38. Another significant page turn (or "hyper-gutter"), from 32 to 33, enacts a "match-cut" between Kudryavka locked up in a dark closet and—page turn!—Korolev having a nightmare about his past imprisonment and waking from troubled sleep in his own unlit bedroom.

39. See Hilary Chute, *Graphic Women: Life Narrative and Contemporary Comics*. Gender and Culture. (New York, NHY: Columbia University Press, 2010), 5 and Hatfield: Chapter 2.

40. Abadzis, *Laika*, 42.

41. Ibid., 72.

42. Ibid., 87.

43. Ibid., 99.

44. Ibid., 100.

45. Ibid., 154.

46. J.M. Coetzee and Amy Gutmann, *The Lives of Animals*. University Center for Human Values Series (Princeton, NJ: Princeton University Press, 1999), 25.

47. David Herman, "Storyworld/Umwelt: Nonhuman Experiences in Graphic Narratives." *SubStance*. 40, No. 1 (2011): 178.

48. As Abadzis said in a 2011 interview: Cartoonists have a tendency to make cute little anthropomorphic characters based on animals we all do it, I do it—which can endow them with the human qualities we might wish they had. But real animals aren't like that, they don't speak, and so we have to remember to speak for them and in a responsible way. Perhaps, speaking from a cultural perspective, we need to anthropomorphize them less, or create more stories and representations that try to respect them as their own sorts of creature, rather than indulge this tendency to humanize them. ... It would help with the way we think about animals as part of human culture. In Coleen Mondor and Lisa Brown, "Animals in Space," *Antennae*. No. 16 (Spring, 2011): 15.

49. Kreiner, "Laika," 190.

50. Brown, "Graphic Novel," 293.

51. , Nelson, "The Legacy of Laika," 210.

52. Abadzis, 51, 55, 78, 175 and passim.

53. Ibid., 166.

54. Turkina *et al.*, *Soviet Space Dogs*, 131.

55. Abadzis, 176–186. Such "experiential frames" appear throughout the novel. See, for example, 125, where Laika appears in a panel overlaid on a widened gutter, with the caption "Good dog," just as Dubrovsky finds out the dog has wound up on the short list for the orbital mission.

56. David Herman, "Storyworld/ Umwelt: Nonhuman Experiences in Graphic Narratives," *SubStance*. 40, no. 1 (2011): 159. As Herman describes its ethical stakes: "[T]he overall impact of the sequence arises from the way it narrates what can happen when the quality or specificity of nonhuman experiences is undervalued or ignored." (158).

57. Elisabeth Bronfen, *Over Her Dead Body: Death, Femininity, and the Aesthetic* (New York, NY: Routledge, 1992), 44.

58. Ibid., 45.

59. Thierry Groenstein, Bart Beaty, and Nick Nguyen, *The System of Comics* (Jackson, MS: University Press of Mississippi, 2007), 45.

60. Nearly the same color as another panel above, which shows Laika floating towards a white light with a lavender halo. In this panel we last see Laika "alive," as a dog, before she appears as only a silhouette.

61. As Judith Butler describes it: "If I lose you … then I not only mourn the loss, but I become inscrutable to myself. Who 'am' I, without you? When we lose some of these ties by which we are constituted, we do not know who we are or what to do. On one level, I think I have lost 'you' only to discover that 'I' have gone missing as well"; in *Precarious Life: The Powers of Mourning and Violence* (London: Verso, 2006), 22. Julia Kristeva says something similar: "[A]ny loss entails the loss of my being—of Being itself. The depressed person is a radical, sullen atheist." In *Black Sun: Depression and Melancholia* (New York, NY: Columbia University Press, 1989), 5. Certainly after losing Kudryavka Dubrovsky becomes just such a "walking dead" sullen atheist.

62. Abadzis, 199.

63. Kreiner, "Laika,"193. Abadzis' use of the photo recalls similar strategies in memoir comics, such as Art Spiegelman's *Maus;* in her analysis Marianne Hirsch relates the move to her concept of postmemory. The animated film *Waltz with Bashir* (directed Ari Folman, Israel, 2008) also closes with live-action footage, to devastating effect.

64. The blog has since been taken down and no direct link exists.

65. R. C. Harvey reports that the public reacted with dismay to Gold's shocking death, with switchboards at major metropolitan newspapers "clogged for hours"—a not-improbable claim in reference to comics' golden age, when the daily strips truly dominated as a mass medium: See *The Art of the Funnies: An Aesthetic History* (Jackson, MS: University Press of Mississippi, 1994), 60. See also Harvey, "Death in the Funnies," *The Comic Blogs Journal* (July 30, 2010) for other examples of death in mainstream comics. We see a more recent, relevant, and very self-reflexive example of an "alternate ending" in Alan Moore and Eddie Campbell's graphic novel *From Hell* (1999), in which a historical victim of Jack the Ripper, Mary Jane Kelly, escapes the killer's clutches to live out her life in Ireland.

66. Unlike in most daily comics, Johnston has allowed her characters to age more or less at a rate comparable to real life; by that measure Farley was nearing the end of his natural lifespan, prompting the artist's choice to give him a "heroic" end. (Spadafori "Pets").

67. Quoted in Spadafori.

68. Excerpts from this letter and others on the Farley case were furnished to me by Caitlin McGurk, from the Lynn Johnston archives at the Billy Ireland Cartoon Library and Museum, Ohio State University. The archives contain roughly 2,000 letters pertaining to the death of the dog.

69. Lynn Johnston, *Remembering Farley: A Tribute to the Life of Our Favorite Cartoon Dog* (Kansas City, MO: Andrews and McMeel, 1996), 189, ellipses in original.

70. Abadzis took the opportunity to create the new strips for the Washington, DC-area comics store Big Planet Comics' 25th anniversary. See Khouri.

71. Abadzis varies the style of each of these new endings, from very similar to his work in the graphic novel to sci-fi psychedelic to garish turn-of-the-century superhero aesthetic, all gritted teeth and erupting "power lines."

72. Abadzis, email message to author April 2016.

73. Ibid.

74. One may access Abadzis' "fifth ending" via this link: http://www.bigplanet-comics.com/wp/wp-content/uploads/2011/08/Laika-alt-05-final.jpg. This short work has more than a bit in common with Moore/Campbell's epilogue to *From Hell*, "The Flight of Gull-Catchers," on the novel and historiography, as well with Spiegelman's rumination on the moral complexities of his self-reflexive approach to the Holocaust at the beginning of part 2 of *Maus*.

75. Such a setting recalls children's books with similar Laika adventures; see Turkina, *et al.*, *Soviet Space Dogs*, 136–137.

76. Abadzis continues to interact with fans on the best way to honor the memory of Laika. As he told me:[T]o this day, I still receive emails and comments online about the ending being so sad, from children and adults of all ages. I greatly appreciate these, and people taking the time to tell me that they read and enjoyed—if that's the right word—my book. I take time to answer every one of them. These days, a link to the alternative endings is provided, which sometimes does seem to provide some salve to readers' feelings. ... I suppose there is a part of me that also did want to say to readers that I understood their emotions. I totally get it. The book's existence is evidence enough, I would say, about how the story affected me as a kid (Abadzis, email message to author, April 16, 2016).

77. Butler, *Precarious Life*, 21; emphasis in original.

78. James Stanescu, "Species Trouble: Judith Butler, Mourning, and the Precarious Lives of Animals," *Hypatia*. 27, No. 3 (2012): 569.

79. Ibid.,577; Elsewhere Butler calls these ties "largely unwilled interdependencies, which imply that the 'ontology' of the human is not separable from the 'ontology' of the animal": *Frames of War: When is Life Grievable?* (London: Verso, 2009), 75–76. Derrida pursues a related idea on the animal's suffering: "Mortality resides there, as the most radical means of thinking the finitude that we share with animals, the mortality that belongs to the very finitude of life, to the experience of compassion, to the possibility of sharing the possibility of this nonpower ..." (28). Abadzis' graphic narrative strategies, as I have argued, model that sharing of the animal's "nonpower" to a significant degree.

80. Becky Ferreira. "Why We Still Want Laika the Space Dog to Come Home," *Motherboard* (November 3, 2014), http://motherboard.vice.com/read/why-we-still-want-laika-the-space-dog-to-come-home.

81. As Abadzis explained in an interview, their centrality even affected to his own revisionist efforts: "That was part of the impulse for creating the book in the first place – some basic desire to put things right, somehow. To give her an escape hatch, a parachute, a way out. Of course, if I was going to remain true to history, I couldn't possibly do that so, in some ways, it was quite a harrowing experience researching and telling the story of her mission,": Mondor and Brown, "Animals in Space": 27.

82. Turkina *et al.*, *Soviet Space Dogs*, 94.

83. Coetzee, Lives of Animals, 63.
84. Ibid., 81–82.
85. Helen Macdonald, *H is for Hawk* (London: Jonathan Cape, 2014), 199.
86. Nelson, "Legacy of Laika," 218.
87. Chloe Taylor, "The Precarious Lives of Animals: Butler, Coetzee, and Animal Ethics," *Philosophy Today.* 52, No. 1 (2008): 64. On animal death in film, see especially Burt, *Animals in Film*, Chapter 3.
88. Colleen Boggs, *Animalia Americana: Animal Representations and Biopolitical Subjectivity* (New York, NY: Columbia UP, 2013), 142.
89. Even though two other dogs, Albina and Mukha, have been established as also present in the room, Abadzis does not depict them at all during Korolev's monologue. To him they are literally invisible, doubly dead.
90. Abadzis, *Laika*, 129–130.
91. Ibid., 131.
92. Ibid., 131–132.

WORKS CITED

Abadzis, Nick. E-mail Interview. April 16, 2016.
———. *Laika.* 1st ed. New York, NY: First Second, 2007.
Alaniz, José. *Death, Disability and the Superhero: The Silver Age and Beyond.* Jackson, MS: University Press of Mississippi, 2014.
Baker, Steve. *Picturing the Beast: Animals, Identity, and Representation.* Manchester, NY: Manchester University Press, 1993.
Berger, John. *About Looking.* 1st Vintage International ed. New York, NY: Vintage International, 1991.
Bezan, Sarah. "Necro-Eco: The Ecology of Death in Jim Crace's Being Dead." *Mosaic: a Journal for the Interdisciplinary Study of Literature.* 48, No. 3 (September, 2015): 191–207.
Boggs, Colleen Glenney. *Animalia Americana: Animal Representations and Biopolitical Subjectivity.* New York, NY: Columbia UP, 2013.
Bronfen, Elisabeth. *Over Her Dead Body: Death, Femininity, and the Aesthetic.* New York, NY: Routledge, 1992.
Brown, Lisa. "An Introduction to the Illustrated Animal." *Antennae.* No. 16 (Spring, 2011): 3–6.
———."A Graphic Novel Raises Ethical Issues." *Society and Animals.* No. 16 (2008): 293–296.
Burt, Jonathan. *Animals in Film.* London: Reaktion, 2002.
Butler, Judith. *Frames of War: When Is Life Grievable?* New York, NY: Verso, 2009.
———. *Precarious Life: The Powers of Mourning and Violence.* New York, NY: Verso, 2006.
Chaney, Michael A. "Animal Subjects of the Graphic Novel." *College Literature*38, No. 3 (2011): 129–49.

———. "The Animal Witness of the Rwandan Genocide." *Graphic Subjects: Critical Essays on Autobiography and Graphic Novels.* Ed. Chaney, 93–100. Wisconsin Studies in Autobiography. Madison, WI: University of Wisconsin Press, 2011.

Chute, Hillary L. *Graphic Women: Life Narrative and Contemporary Comics.* Gender and Culture. New York, NY: Columbia University Press, 2010.

Coetzee, J. M., and Amy Gutman. *The Lives of Animals.* University Center for Human Values Series. Princeton, NJ: Princeton University Press, 1999.

Derrida, Jacques. *The Animal That Therefore I Am.* Edited by Marie-Louise Mallet. Translated by David Wills. New York, NY: Fordham University Press, 2008.

Ferreira, Becky. "Why We Still Want Laika the Space Dog To Come Home." *Motherboard* (November 3, 2014). http://motherboard.vice.com/read/why-we-still-want-laika-the-space-dog-to-come-home.

Gardner, Jared. "Sidney Smith's Open Road." *The Library of American Comics Essentials, Vol. 2: The Gumps by Sidney Smith.* Ed. Gardner, 7–25. San Diego, CA: IDW Publishing, 2013.

Gomes, Mario and Jan Peuckert. "Memento Mori: A Portuguese Style of Melancholy." *Comics as a Nexus of Cultures: Essays on the Interplay of Media, Disciplines and International Perspectives.* Ed. Mark Berninger *et al.*, 116–126. Critical Explorations in Science Fiction and Fantasy; 22. Jefferson, NC: McFarland, 2010.

Groensteen, Thierry, Bart Beaty, and Nick Nguyen.*The System of Comics.* Jackson, MS: University Press of Mississippi, 2007.

Harvey, Robert C. "Death in the Funnies" *The Comics Journal Blog* (July 30, 2010). http://classic.tcj.com/strips/death-in-the-funnies/.

———. *The Art of the Funnies: An Aesthetic History.* Studies in Popular Culture (Jackson, Miss). Jackson, MS: University Press of Mississippi, 1994.

Hatfield, Charles. *Alternative Comics: An Emerging Literature.* Jackson, MS: University Press of Mississippi, 2005.

Herman, David. "Storyworld/Umwelt: Nonhuman Experiences in Graphic Narratives." *SubStance* 40, No. 1 (2011): 156–81.

Hirsch, Marianne. "Mourning and Postmemory." *Graphic Subjects Critical Essays on Autobiography and Graphic Novels.* Wisconsin Studies in Autobiography. Ed. Chaney, Michael A. Madison, 17–44. Madison, WI: University of Wisconsin Press, 2011.

Johnston, Lynn. *Remembering Farley: A Tribute to the Life of Our Favorite Cartoon Dog.* Kansas City, MO: Andrews and McMeel, 1996.

Khouri, Andy. "Laika the Space Dog Survives in Alternate Happy Endings by Nick Abadzis." *Comics Alliance* (August 8, 2011). http://comicsalliance.com/laika-happy-endings-nick-abadzis/.

Kreiner, Rich. "Laika." *The Comics Journal.* No. 292 (October, 2008):190–193.

Kristeva, Julia. *Black Sun: Depression and Melancholia.* European Perspectives. New York, NY: Columbia University Press, 1989.

Macdonald, Helen. *H is for Hawk.* London: Jonathan Cape, 2014.

McCloud, Scott. *Understanding Comics.* Northampton, MA: Kitchen Sink, 1993.

Mondor, Coleen and Lisa Brown. "Animals in Space." *Antennae.* No. 16 (Spring, 2011): 8–28.

Nelson, Amy. "The Legacy of Laika: Celebrity, Sacrifice and the Soviet Space Dogs." *Beastly Natures: Animals, Humans, and the Study of History*. Ed. Dorothee Brantz, 214–224. Charlottesville, VA: University of Virginia Press, 2010.

———. "The Music of Memory and Forgetting: Global Echoes of Sputnik II." *Remembering the Space Age*. NASA SP (Series); 4703. Ed. Steven J. Dick, 239–250. Washington, DC: National Aeronautics and Space Administration, Office of External Relations, History Division, 2008.

Spadafori, Gina. "Pets a Big Part of the Cartoon Universe." *Sacramento Bee*. October 29, 1995. http://www.spokesman.com/stories/1995/oct/29/pets-a-big-part-of-the-cartoon-universe/.

Spurgeon, Tom. "A Short Interview with Nick Abadzis."*The Comics Reporter* (September 1, 2007). http://www.comicsreporter.com/index.php/resources/interviews/9884/.

Stanescu, James. "Species Trouble: Judith Butler, Mourning, and the Precarious Lives of Animals." *Hypatia* 27, No. 3 (2012): 567–582.

Taylor, Chloë. "The Precarious Lives of Animals: Butler, Coetzee, and Animal Ethics." *Philosophy Today* 52, No. 1 (2008): 60–72.

Turkina, Olessya, Inna Cannon, and Lisa Wasserman. *Soviet Space Dogs*. London: FUEL Design & Publishing, 2014.

Verdery, Katherine. *The Political Lives of Dead Bodies: Reburial and Postsocialist Change*. Annual W. Averell Harriman Lecture. New York: Columbia University Press, 1999.

Chapter 4

The Anterior Animal

Derrida, Deep Time, and the Immersive Vision of Paleoartist Julius Csotonyi

Sarah Bezan

In his photographic composites and digital paintings, Canadian scientific illustrator Julius Csotonyi presents a hyper-realistic view of primordial life that imbeds his viewers in an immersive capsule of deep time. Vividly rendered in hues that range from saturated tones of indigo and cobalt to ombré shades of burnt orange and chlorophyll-green, Csotonyi's portraits of ancient marine reptiles and extinct terrestrial megafauna chromatically reanimate the sedimented terrains and seascaped vistas of a planet in its earliest throes of life. As a jointly speculative and scientifically accurate restoration of long-dead animal ancestors, these paleontological illustrations signal a divergence from classical paleontological art that depict prehistoric life as a series of static episodes of evolutionary progress that culminate in the teleological emergence of the human species. Alternatively, Csotonyi's work self-reflexively gestures to the methods by which artists create an animate portrait of the past out of the seemingly inert, fragmented matter of the articulated fossil. Through the use of biphasic waterlines, a bead of mercury, and the aqueous humor of a theropod dinosaur eye, Csotonyi's illustrations emphasize the posteriority of the human–that is, the emergence of the human species after the venerable elapse of prehistoric epochs. Analyzing how these visual narratives of evolution come to be contrived and subverted, this chapter asks: to what extent can paleontological illustration make visible that which came before the human—namely, that which is incontrovertibly unobservable?

To launch such an inquiry into an encounter with the anterior animal, we begin with Jacques Derrida's own engagement with anteriority and deep time in *The Animal That Therefore I Am*. Though now a well-known and widely discussed passage in recent scholarly treatises of animality, the scene in which Derrida is seen naked by his cat has not yet been apprehended by scholars as a provocation to thinking about the animal in geological time.

In his ruminative survey of the abyssal limit of the human and the animal, Derrida returns recursively to the question: *who comes before and who is after whom?* To be sure, the problem of anteriority hovers unyieldingly in the foreground of Derrida's thinking. Yet, as I argue in this chapter, scholars of Derrida have overlooked the implications of his reflections on the passive stance of the human, in which we are always already seen by the animal, *arriving after it in a time before time.* The yawning berth of time that separates, and yet also tenuously connects forms of life by succession, remains a shared concern both of Derrida and of thinkers in the emergent sub-fields of new materialism, speculative realism, and object-oriented ontology (despite the intervention of the latter sub-fields into the linguistic turn of which Derrida is a part). Arguably, anteriority is a problem for Derrida and for thinkers of the material turn, as well as for paleontological scientists and paleoartists: as an antecedent of all that we deem to be human, the anterior animal occupies a dizzying and disorienting locus point in time.

Advancing a renewed critique of deep time in Derrida's thought, and further identifying the fossil as a site of common ground between the linguistic and material turns, this chapter examines how Csotonyi's hyper-realistic paleoart formally subverts the primacy of human vision and enables a critique of anteriority. By utilizing immersive and embedded perspectives, indiscernible foregrounds, shifting frames, and expansive scales, Csotonyi carefully composes an image of the paleontological object that tests—and even embraces—the limits of what is possible to validate through human perception. While Csotonyi's illustrations acknowledge the recent emergence of the human along with its intimate entanglement in the ongoing and perpetual production of the history of life on earth, they are also a representation of the past that is conceived in and through the productive tension that exists between what is known and what can be imagined. Moreover, Csotonyi's paleoart is self-critical in its utilization of the visual form, marking a departure from the conventions of what has traditionally subtended the practice of paleontological illustration over the past century. Unlike the majority of paleoart, which largely exhibits prehistoric animals as fixed and definitive prototypes of evolution in its ascent to the supposed supremacy of the human, Csotonyi's illustrations promote a circumspect awareness of the deep anthropocentrism that is inherent to any representation of geological time that blindly idealizes "Nature" as an abstraction of that which falls within the human field of vision. In its self-reflexive critique of the ocular mode, Csotonyi's work troubles the notion that the human wields an unmediated, stable, or superior point of view of primordial life.

If a stabilized perspective on the anterior animal is lacking in our artistic renderings and in our treatment of evolution, it is because the expanse of deep time is ultimately insurmountable and incogitable. Derrida elaborates on the scopious nature of time in his discussion of the Edenic episode of naming, wherein "God thus lets Ish do the calling all alone; he accords him the right to give them names in his own name—but just in order to see."[1] As such, the act of "being *after*, being *alongside*, being *near*"[2] the animal takes place, Derrida writes, in "*this time before time that has always made me dizzy*."[3] It is at this inaugural moment that the abyssal limit opens up into a multiple and heterogeneous border between the living and dead human and nonhuman animal, effectively countering any interpretations of being and time as strictly chronological, historical, or static. Expounding further on the animal question, which explicitly correlates with the intricacy of succession (to *follow*, and to be *after*), Derrida turns to the "bottomless gaze" of his cat.[4] The feline's gaze reflects back upon him the truth of *seeing*, and more importantly, of *being seen*—not through the parameters of historical progress, but within the dizzying emanation of deep time. Through the mammalian gaze of a cat, Derrida gestures to the multimillion-year-old age of ancient reptiles, avian dinosaurs, and other earlier iterations of prehistoric life, leaving his readers to consider how the processes of observation and nominalization in Genesis (the time of the creation of the world) are in fact broken down into a question of precedent. "For so long now," Derrida goes on, "it is as if the cat had been recalling itself and recalling that, recalling me and reminding me of this awful tale of Genesis, without breathing a word. *Who was born first, before the names? Which one saw the other come into place, so long ago?*"[5] In Derrida's vision, the act of following and tracing the animal as an autobiographical gesture is bound up in the prospect of anteriority and deep time.

By interpreting animality through a temporal framework that emphasizes the bountiful interconnections between past/present and the human/nonhuman, Derrida's consideration of time before the human makes an important contribution to ongoing debates about materiality and being in the world. We learn in the latter half of *The Animal That Therefore I Am* that time, which Derrida regards as, "the transcendental horizon of the question of being,"[6] is also the common root of the problems of "world, individuation, and finitude"[7] in the Heideggerian tradition. Contra Heidegger's anthropocentric philosophy of time, however, Derrida hones in on the generativity of the time lapse that spans the moment of creation and the assignation of names. As Joanna Hodge argues in *Derrida on Time*, this moment of creation is proportionately disruptive and multiplicitous: it distinguishes the "time of animals, with a delay between their creation, and their naming," from "the time of the language of tenses, conditionalities and reversibilities."[8] In calling up this ancient scene, Derrida notes the jarring, disorienting, and yet fertile suspensions of time. In short, while Heidegger understands temporality as both a measure and articulation of being, Derrida explicates the time before the human (prior to the observation and nominalization of animal life) as an inexhaustible and untethered relationship between past and present, human and animal. "We no longer," Derrida writes, "even have a clock or chronological measure of it, we, we who call ourselves men, or humans [...] This new situation can be determined only on the basis of what is most ancient."[9] Derrida therefore unequivocally maintains that the heart of the question of animality is located not merely in history or language but in this detonation of time and in the emergence of living things "that came into the world before him but were named after him."[10] Through this retrospective gesture toward the primordial past, Derrida emphasizes how he, much like his readers, is invariably caught in the gaze of the anterior animal and thereby inextricably hinged in the present (*after*) and the past (*before*).

Yet in meditating upon the temporal schism in the tale of Genesis as a manifolding of being in the expanse of prehuman time, Derrida has been accused (and to some extent, rightly so) of eschewing the material reality of nonhuman animals. For instance, in her critique of the "draining of historical materiality out of the sign of animal life,"[11] Nicole Shukin aptly observes that the spectral animal of this scene is framed "in the same terms he uses to frame the ghostly visitation of Hamlet's father" in *Spectres of Marx*, and in so doing, "Derrida risks collapsing the material difference between the body of an actual animal and the prosthetic armor of a fictional specter."[12] This emptying of "historical substance," Shukin argues, means that the "spectral animal visually channels the disquieting half-presence of a "life" never consubstantial with terrestrial Time, History, and Being."[13] Although Derrida leans away from these tenets of Western ontology in his reflections on the animal, I would contend that such an abnegation veritably clears the way for a renewed engagement with materiality. Posthumous life—that is, all human and nonhuman animal forms that precede us—is, as Derrida suggests elsewhere, "becoming the very element, mix[ing] in everywhere with the air we are breathing."[14] Proceeding from this claim, we might reinterpret the anterior animal—or at least its remains in the form of the fossil—as a marker of the *at once hidden* (spectral, absent) *and yet proximate* (material, present) postmortem world that is *after*, *alongside*, and *near* us. In short, the anterior animal lies largely beyond the parameters of our view and outside our conception of historical and evolutionary progress.

So concealed, this postmortem world marks the limits of what we can see, and consequently of what we can know. And by intimating this fact, Derrida's deference to the enigma of deep time can be favorably compared to the key arguments and interests of thinkers of the materialist turn. In radicalizing the "human-world" correlate from within (i.e., the so-called correlationism that pervades post-Kantian philosophical thought), speculative realists Quentin Meillassoux, Graham Harman, and Levi Bryant aim to discharge the human from its role in validating reality through its own perceptive apparatus.[15] In *After Finitude*, Meillassoux discusses his concept of the *arche-fossil* (understood both as geological material and as an "ancestral statement" that indicates the prehistoric presence of organic life), insisting that what is at stake in a critique of correlationism is "the nature of scientific discourse" and its capacity to empirically describe a world that precedes human thought.[16] Similarly, Timothy Morton's commentary of the fossil, read through the lens of object-oriented ontology, reconfigures the paleontological object as *hyperobject* that maintains a "sensuous connection" with the present even as it withdraws its "primordial reality" from humans.[17] Along with Meillassoux and Morton, a wide variety of scholars in the environmental humanities, ecocriticism, posthumanism, and animal studies more broadly are likewise invested in dismantling trancendental humanism. More specifically, scholarship in these fields remains critical of *the human's strategies of mediation of the natural world*, whether it be through consciousness, language, spectrality and auxiliary modes of sensation, or other preemptive methodologies of reason. Combating transcendentalism comes down to finding a way to think through what Meillassoux describes as the "inside/outside" problem,[18] in which the transcendental subject *takes place* and maintains a *point of view*.[19]

This transcendental mediation of "Nature," by and large, forms the backdrop of the better part of paleoartistic renderings of the past century, as I will soon illustrate. Yet Csotonyi's immersive vision, I argue, creatively engages with the "inside/outside" problem of transcendentalism, and further creates new ways of drawing attention to mediation itself. Most compellingly, in "Troodon's Gaze" (Figure 4.1), Csotonyi invites his viewers to consider temporal anteriority and the co-constitutive encounter of the human and animal in deep time. This piece renders the human in and through the prescient optics of the ancient Troodon, a bird-like dinosaur of the late Cretaceous period. Commissioned for the cover of The University of Alberta's Alumni Magazine *New Trails* in support of a story of the university's prestigious paleontological research team, the painting bears the image of paleontologist Scott Persons, who is reflected in, and seemingly installed inside (yet also beguilingly outside), the eye of the Troodon.[20] The ancient avian gaze holds the paleontologist in view, as if anticipating the moment at which the research team (composed in part of Philip Currie, Michael Caldwell, Victoria Arbour, and Scott Persons himself) will dig up its postmortem fragments. Imagining a continuous affinity between inside and outside, and between the living and dead human and nonhuman animal, the frame of the painting captures the colorful Troodon eye (unique because of its visual acuity—a characteristic of theropod dinosaurs),[21] which has in turn already appropriated the trace image of the paleontologist. The circuity of the gaze (between the observer and observed, human and avian) is coupled with the emanation of the Troodon's own binocular vision, which overlaps fields of view: in this

Figure 4.1 ***Troodon's Gaze.*** Digital painting and photographic composite. Julius Csotonyi, 2013.

case, the foreground (the past) and the background (the present). Csotonyi's painting of the double vision of the Troodon and the paleontologist in deep time invites us to ask: *Who is first? Who is looking at whom?*

Such an illustration of anteriority can be contrasted with the comparative anatomical drawings and life-like sculptures of early hominids by paleoanthropological artists like John Gurche, Artist in Residence at the Museum of the Earth in Ithaca, New York, and author of *Shaping Humanity: How Science, Art and Imagination Help Us Understand Our Origins* (a collection based on his contributions to the Smithsonian Institution's National Museum of Natural History's Hall of Human Origins). Imbued with a glimmer of contemplative consciousness—the surmised sensibility that is thought to be an essence of the modern *Homo sapiens*, Gurche's realist sculptures notably feature the human's evolutionary development of the pale sclera (the white of the eye), a distinguishing aspect of human vision that nonverbally communicates which direction the hominid is looking. From the *Australopithecus afarensis* (commonly known as "Lucy") to the *Homo erectus*, Gurche's contributions to the Smithsonian furnish a narrative of the human that tracks the nonverbal communication of early hominids and privileges visuality as one of the markers of higher intelligence. Endowing these extinct humans with the power to look (and even to ostensibly look back at the modern human observer or museum-goer), Gurche's work correlates the origin of the human with the advancement of the hominid's visual apparatus.

Along with Gurche, Viktor Deak, co-author with physical anthropologist Gary J. Sawyer of *The Last Human: A Guide to Twenty-Two Species of Extinct Humans* and contributor to the American Museum of Natural History's Anne and Bernard Spitzer Hall of Human Origins, is well known for turning "fossils into faces"[22] using the most recent paleoanthropological research of human evolution. Deak's work too notably incorporates the white of the eye into some of the later species of extinct humans, such as the *Homo rudolfensis*.[23] Although Deak willingly concedes that the origin of the human is uneasily determined,[24] it is nevertheless the case that art by paleoanthropological artists like Gurche and Deak often maintains an exclusive focus on the hominid based on the notion that there is a conclusive and discernible break between the great ape and the human, and between the varieties of extinct hominid species that are endowed with the seemingly superior capacity to visually perceive and interpret the environment and its inhabitants. While as much a creative and interpretative endeavor as any other paleoartistic project, the representation of the increasingly observant (and therefore more "human") hominid in paleoanthropological art comes to reflect the very transcendentalism that materialist scholarship would critique.

Following this transcendentalist schema, the idea of nature that grounds much of paleoanthropology research and artistic expression comes down to a strict and unrelenting limit of what is, and what is not, human. Jessica Mordsley astutely observes that the paleoanthropological division of *Homo* and *Australopithecus* fossils seems to mark "a clear dividing line between prehuman animal and human," but notes that "this dividing line is both arbitrary and disputed."[25]A deeply humanistic science, Mordsley argues, paleoanthropology depends on the slippery signifier of the human[26] which is always constructed based on a notion of what sets the human apart from its ancestors. In her salient critique of the aims of paleoanthropology, Mordsley turns to Jacques Derrida's paradigm of *différance* and to Charles Darwin's frequent references to "the living trace." Here, Mordsley analyzes the play of traces that cross "the limits of human/animal, present/past, living/dead," which in turn suggests, she argues, that "*the human is never simply present or absent*."[27] By rearticulating the complex process of variation and differentiation between species, Mordsley insists that all living things are rooted in ancestral connections with earlier life forms: there are, she writes, parts of our predecessors alive in us all.[28] Uniting time, being, and place, Mordsley's observations echo Eileen Joy's "You Are Here" manifesto on the ethics and agencies of objects, in which she argues that "the human body is itself a time capsule of all previous bodies … . Nothing is ever lost, although if [Graham] Harman is right, everything is always withdrawing from everything else. Again: you are here, but a part of you is also somewhere else."[29] Both present and absent, materially visible and withdrawn, the necessarily incomplete fossil record cannot empirically recount these fleshly ancestral associations and connections in deep time. To interpret the play of traces among species, one must first assume that the lion's share of evolutionary intonations is largely unknown and unseen.

Understanding Derrida's abyssal limit as an admonition of deep time, and decoding the fossil as a material marker of the intersection of the "human/animal, present/past, living/dead" (as Mordsley delineates), we can begin to liberate meaning from matter, or, to put it more succinctly, to presume the inherent meaning of matter independent of the human observer. Thus, following scholars of the materialist turn who are continually reworking Derridean deconstruction and poststructuralist principles into their thinking of a nonanthropocentric paradigm of matter and meaning—from Karen Barad's "agential realism" and Wendy Wheeler's "biosemiotic realism" to Serenella Iovino and Serpil Opperman's material-ecocritical analysis of "storied matter"—we can come to acknowledge the varied ways in which the world is endowed with a narrative agency and a "more-than-human materiality"[30] that positions the human not as an outside observer but as an immersed and embedded body: "a part of the world in its ongoing intra-activity."[31] Csotonyi's own immersive vision in "Troodon's Gaze" reflects the ethos of this nonanthropocentric materialism, reconstructing the human inside and within the world, across deep time, and beyond the bounds of the life-death distinction.

The implications of this immersive perspective become clearer when we situate Csotonyi's art in the wider context of the history of paleontological illustration. To do so, we must begin with the process by which postmortem matter (i.e., the fossil) comes to be revivified in the semblance of life and featured in paleoartistic illustration. We learn from renowned Cambridge historian Martin Rudwick, for example, that from the inception of paleontology as an established scientific discipline in the early nineteenth century, the fossil was simply understood as a thing (whether organic or inorganic) that is "dug up."[32] In other words, *the definition of a fossil is necessarily hinged on a human observer—trowel and chisel in hand.* This then means that the abundant remains of prehistoric organisms that presumably lie undiscovered and embedded in the earth's vast layers of strata are not, properly speaking, fossils. It also means that the definition of the fossil cannot account for the immense and incalculable volume of dead organisms whose remains were scavenged, weathered, decomposed, recreated into new material configurations or otherwise irretrievably lost forever in ways that the human can never discern. Whatever fossils we do discover, dig up, and examine are therefore a small, static, and inert snapshot of an intensive process of variation across a vast period of time. So how do we think through and represent these unobservable gaps in the fossil record?

Henry De la Beche's 1830 watercolor painting *Duria Antiquior, A More Ancient Dorset*, a much-copied and lithographed piece created in response to the discoveries of English fossil collector and paleontologist Mary Anning (see Figure 4.2), serves as a prime example of the recuperative representation of prehistoric life. At this time in the study of paleontology, it was only becoming more widely acknowledged, as Donald Grayson notes, that fossils were not merely "objects that had grown in place in the earth, but were instead the remains of actual, once-living animals now buried beneath the surface."[33] Following French naturalist Georges-Louis Leclerc Buffon's gargantuan thirty-six volume encyclopedia *Histoire Naturelle* (1749–1804), which conceived of the ancient past of the earth in the context of the Mosaic chronology, along with comparative anatomist and naturalist Georges Cuvier's drawings of the megatherium and other extinct species, an increasing

Figure 4.2 ***Duria Antiquior, A More Ancient Dorset.*** Watercolor painting. Henry De la Beche, 1830.

number of naturalists began to rebuild the history of life on earth using the evidence of fossils. De la Beche's *Duria Antiquior* was the first widely disseminated and commercialized piece of artwork that explicitly demonstrated this newfound sense of ancient history. As Jane Davidson writes in *A History of Paleontological Illustration*, De la Beche's depiction of ammonites, fish fossils, paleobotanicals, pterodactyls, ichthyosaurs, and plesiosaurs made *Duria Antiquior* a veritable "compendium of much of what was known about fossils in 1830."[34]

Edward Drinker Cope, Samuel Wendell Williston, and a number of other paleoartists collaborating closely with scientific researchers (or working as palaeontologists in their own right) closely followed De la Beche, each in turn creating their own imagery and perspectives on the deep past. Some, Davidson explains, were less accurate than others in representing the paleontological animal as it would have evidently appeared in life. Davidson thereby concludes that the aesthetic conventions shared by many paleontological illustrations from the fifteenth century to the present include some elements of fantasy, regardless of whether or not this was the intention of the scientists and artists in their collaborations.[35] In addition, what most united paleontological illustrators according to Davidson was their energetic experimentations with new media,[36] from the use of pastel, watercolor, woodcuts, copperplate engravings, etchings, lithography, crayon, pencil, chalk, sculpture, and clay modeling, to more recent innovations in moving pictures, cartoons, animatronics, animation, computerization, digital painting, photorealism, composite photography, murals, museum dioramas and exhibits, and even merchandise. From the realistic woodcuts of the baroque artists during the fifteenth and sixteenth centuries to the sculptures of ichthyosaurs, plesiosaurs, and other extinct species by Benjamin Hawkins Waterhouse in London's Crystal Palace Park (unveiled in 1854 under the scientific advisement of Sir Richard Owen), representations of the past world of extinct animals were designed to show evidence of scale, perspective, and the primitive animal's relationship to its ecological settings.

As Stephen Jay Gould writes in the October 1993 issue of *Discover*, the rise of paleontological art depicting dinosaurs was also immensely influenced by powerhouse painter Charles R. Knight (1874–1953), whose technically brilliant and naturalistic mural paintings came to be housed in the Natural History Museum of Los Angeles County, the Chicago Field Museum of Natural History, the Smithsonian, the American Museum of Natural History's Dinosaur Hall, and numerous other distinguished institutions of paleontological research. Over the duration of his long career (during which his eyesight increasingly failed, despite his legacy as "the artist who saw through time"), Knight brought to life countless fossils that had only before been exhibited "as they appeared in nature," often still encased in "the surrounding rock or matrix."[37] In his description of Knight as "the great iconographer of dinosaurs," Gould illustrates how Knight's drawings, clay models, sculptures, and murals came to define the genre of paleontological art at the turn of the twentieth century. Gould narrows Knight's influence down to three "defensible" conventions: *number*, *activity*, and *emphasis*.[38] Due to the limited space allotted for murals, Gould writes, paleoartists had to make-do with the limited space: "If I am allowed but one picture of a Mesozoic landscape, I will therefore try to wedge everything in— predator and prey, pond dweller and mountain climber."[39] Similarly, Gould suggests that paleoartists often depicted an array of primitive animal behaviors, echoing De la Beche's famous work in its "conventions of unnatural crowding and pervasive predation."[40] Lastly, in a critique of the narrowness of paleoart to date, Gould laments the limited scope of paleoart in its over-representation of vertebrates: this emphasis, he writes, means that "most of life's history gets scrunched into an introductory picture or two."[41]

What is most compelling about Gould's analysis of Knight's work and the conventions of paleoart (following his critique of the convention of emphasis) is his assessment of the paradoxical problem of portraying the history of life in short slivers. Confronted with the principle of contingency, which rules that the continuous narrative of the history of life is governed by chance, the limited view on the past merely condenses and overshadows the conditional variables that have influenced evolutionary processes. As Foucault observes in his discussion of fossils, natural history proceeds not by representing the shifting contingencies of the history of life but through a *paradigm of continuity that builds a reliable prototype*: by "integrating the series of successions with the continuity of beings and their distribution over the table," creatures can thereby be taxonomized in the manner of a "progressive hierarchization."[42] Given the complexity of deep time and the tendency to which natural history flattens itinerant waves of variation in the evolution of species, Gould argues that, "we do not even know how to conceptualize, much less to draw, the worldview that would place *Homo sapiens* into proper relationship with the history of life. We know the iconography of directional pageantry, for we have been drawing history according to this scheme for centuries, but what is the iconography of contingency?"[43] What Gould identifies is the perennial problem facing paleoartists: how do you pin down the ever-shifting processes of life's ongoing expression?

Natural history museums across the Western world, while increasingly more diligent and attentive to this problem, are nevertheless steeped in traditions that belie an old allegiance to the regime of order, classification, nominalization, and spectacularization in their representation of the natural world. From the genealogical series patterning of horse species in the American

Natural History Museum (analyzed by Jeremiah Dyehouse)[44] to the spectacular appearance of the famous and imposing "Mahajarah," an African elephant housed at the Manchester Museum (as described by Samuel Alberti),[45] the concepts of nature that remain in circulation indicate a lingering presence of the Great Chain of Being. Much like the acrid smell of the bactericides and oils applied to many of the aging taxidermy displays, what persists is a hint of the once-prevalent narrative of the teleological advancement of the human that draws together the history of life into tidy exhibits.

The persistence of these traditions can still be observed in more contemporary paleoart. The diorama of the ancient mosasaur painted by Karen Carr (housed at the Canadian Fossil Discovery Centre in Morden, Manitoba, Canada), for instance, exemplifies a mimetic or mirror image of the fossil (see Figure 4.3). In the exhibit (which is enclosed in glass and accompanied by a placard indicating its name and item number), we see the foregrounded mosasaur fossil mirrored in Carr's painting, represented in the pretense of life. Elsewhere in the Canadian Fossil Discovery Centre, a fully articulated mosasaur skeleton model, "Suzy," is displayed alongside a large, colorful mural that similarly parallels the massive specimen (see Figure 4.4). In both

Figure 4.3 "Mosasaur Skull." Museum diorama by Karen Carr. Canadian Fossil Discovery Centre. Photograph by the author. (date unknown).

Figure 4.4 **"Suzy."** Mural artist unknown. Canadian Fossil Discovery Centre. Photograph by the author (2016).

of these exhibits, the mimetic composition of the paleoart makes the fossil a "privileged locus of a resemblance," in Foucault's analysis.[46] In doing so, both the mural and the paleoart by Carr reveal their adherence to the traditional naturalist program of nominalizing and structuring the visual world through the principle of continuity. In this paradigm, museum-goers merely confirm what they expect to see: the prototype is copied infinitely, flattening the dynamic history of life's ongoing expression in deep time onto a painted exhibit wall.

What sets Csotonyi apart from other artistic work that represents the paleontological animal as a flat surface, mirror image, or a static prototype is his inventive focus on perspective in his digital painting and photography. Along with his (re)vision of the paleontologist inside and within the gaze of the ancient Troodon, Csotonyi's work plays with frames, scale, and point of view in order to challenge the viewer's position. In his masterful digital painting, "*Apatosaurus* Tree Tipping" (Figure 4.5), Csotonyi reorients and reshapes a conventional panoramic piece into what he describes as a "spherical mirror." The idea for this piece, Csotonyi writes in his book, arose from a conversation with paleontologist Dr. Dave Hone of Queen Mary, University of London, at a Fossil Preparation and Collections Symposium hosted by the Royal Tyrell

Figure 4.5 "Apatosaurus Tree Tipping." Digital Painting. Julius Csotonyi. 2013.

Museum in Drumheller, Alberta. In their tour of the Royal Tyrell museum's exhibits, Hone and Csotonyi found themselves discussing the *Apatosaurus* and its destructive feeding habits (tree-tipping to feed on upper branches).[47] As a result of their discussion, Csotonyi decides to challenge the viewer's expectations of the long-necked sauropod, which is depicted in paleoartist John Sibbick's rendition from an aerial view. The deliberate distortion of views in "*Apatosaurus* Tree Tipping" reflects Csotonyi's aim to place the viewer inside "a bead of mercury," thereby destabilizing point of view and embedding the observer inside a fluid capsule.[48] Emphasizing the nonlinear outline of the sauropod itself, the circular frame of the piece reveals the massive scale of the animal in relation to the verdant blades of grass that surround the sphere on the ground. Inside this bead of mercury, the viewer is positioned as if "sitting on the top of the stem of a horsetail plant in the Jurassic landscape that generated the Morrison Formation."[49] The mercurial fluidity (and minute scale) of the viewer's point of view creates a remarkable image of the colossal *Apatosaurus*, working to imaginatively embed the viewer in the earth as its own kind of fragmented fossil. In its distorted multidimensionality, the piece does not settle on a single limit or a horizontal line but is instead oriented around a zenith position that is connected by a series of curved lines. It is a remarkable image of the paleontological animal that renegotiates the bounds of human perception.

In another collaboration with Dr. Dave Hone, Csotonyi's digital painting, "*Rhamphorhynchus* Gleaning Squid" (Figure 4.6) visually represents Hone's paleontological research on the feeding habits of the small Jurassic pterosaur *Rhamphorhynchus*. As Hone explains in a caption to the image in Csotonyi's book, the *Rhamphorhynchus* is believed to have preyed on squid, based on findings of their fossilized coprolites (feces).[50] In artistically representing this research, Csotonyi composes a doubled biphasic image that positions the viewer as a bystander from an underwater perspective. The goal, Csotonyi writes, was "to clearly show both animals when they are separated by the visually distorting interface between water and air." In comparison to De la Beche's hard horizontal separation of the waterline in *Duria Antiquior*, Csotonyi's digital painting utilizes the effects of water distortion in order to represent both predator (*Rhamphorhynchus*) and prey (squid). Csotonyi describes his process of using light and refraction to compose his image:

> As a rule, I block in the background first and work on the focal subject later, because the nature of the light that plays on the subject is much more reliably

Figure 4.6 "Rhamphorhynchus Gleaning Squid." Digital Painting. Julius Csotonyi. 2013.

> determined once I establish the locations of the complex reflective, refractive, and occulting surfaces of the environment that generate sources of illumination and shadow. Understanding the physics of light, especially the processes of reflection, refraction, diffraction, scattering, interference, black body radiation, and absorbance spectra of materials, will take an artist a large proportion of the way to a successful painting.[51]

Following all of the criteria set out by Snell's law (the law of refraction between water and air), the painting is both scientifically admissible and imaginatively constructed. As the mandible of the pterosaur cuts into the surface of the water, the viewer is immersed in an underwater perspective that offers a close-up, and yet aqueous (and therefore naturally distorted) perspective of the primordial animal. Csotonyi's meticulous reproduction of the effects of natural elements and human vision is indicative of his own painstaking attention to scientific principles, but also demonstrates the limits of our vision. By enclosing the human observer in the water and earth of the primordial planet (through the aqueous humor of the eye, the bead of mercury, and the biphasic waterline), Csotonyi's immersive art perhaps more accurately reflects our own dizziness at the prospect of encountering anterior life. This is the vertiginous *time before time* that displaces our ontological position and shifts our epistemological footing.

Plunging into this whirling *time before time*, outside the bounds of the natural history museum and beyond the glossy pages of Csotonyi's coffee table book, reverberations of Derrida's session on the anteriority of the animal continue to echo in my brain. The flat and perfunctory mural of the mosasaur that appears upon the walls of the Canadian Fossil Discovery Centre implies that all of time is but a succession of uncomplicated evolutionary productions. *But where do we fit, given our posteriority, in this ever-unfolding narrative of life?*

As I learned on a recent paleontological dig, an actual encounter with the anterior animal is unsteadying. Unlike the subdued, one-dimensional, and mimetic images of the mosasaur, it was the exposed turf of a dig site that had not seen the light of day in eighty-million years that precipitated the crumbling sediment of my own thinking about prehistoric life. What was once, unfathomably, the watery expanse of the Western Interior Seaway is now the fertile farmland of the prairie landscape in Manitoba, Canada. The seabed is an archive of ancient aquatic life, from sea turtles and microorganisms to the massive vertebrate reptiles like the plesiosaur and mosasaur: it is the very scene of Derrida's unthinkable and unobservable "abyssal limit." On a clear, sunny day in May 2016, I joined paleontology tour guide Nolan Turnbull for an afternoon of digging at an excavated site located midway down a crevasse lined with shale, concretions of microfossils, and fossilized fish fragments

Figure 4.7 "Tarping the Dig Site of the Mosasaur." Pembina Escarpment, Manitoba, Canada. Photograph of the author (2016).

(see Figure 4.7). Together, we unearthed nine vertebrae belonging to a mosasaur in the layers of sediment. Caked in dust, and equipped with a set of dental implements for carefully extricating the fragments of what was once likely a thirty to forty-foot carnivorous marine reptile, I apprehended the limits of what I could see and know of the death of this creature that lived so long ago, before me. Before, alongside, and near the anterior animal, I was dizzy. At one moment stationed as a kind of transcendental observer: one who *takes place* and maintains a *point of view* (in the words of Meillassoux), I instead became disinterred. I was, myself, unearthed and exhumed under the gaze of the awaiting animal (see Figure 4.8). I learned that once confronted with the stark, incomprehensible materiality of prehistoric life, it becomes possible to reflect on how we might resist relying on the abstract ontological principle about the natural world that, as Morton would say, "transcends the material realm or even the realm of representation."[52] As I have argued, Csotonyi's paleoart traverses the disciplines of paleontological science and digital artistry, and his retreat into primitive natural history is both palpable and (conceivably) realist. Like Derrida, who imagines a time before the observation and nominalization of the animal, Csotonyi illustrates a world in which the absent human is an anticipated expression of life's evolutionary unfolding: a proverbial twinkle in the anterior animal's eye.

Figure 4.8 "Mosasaur Vertebra." Pembina Escarpment, Manitoba, Canada. Photograph of the author (2016).

ACKNOWLEDGMENTS

I am very grateful to Dianne Chisholm (University of Alberta) and Katja Herges (University of California, Davis) for their insightful comments on an earlier draft of this chapter, and to The University of Manitoba Institute for Humanities for inviting me to share this research at their Research Colloquium. Furthermore, I thank Nolan Turnbull and the rest of the staff of the Canadian Fossil Discovery Centre for setting the stage for my encounter with a mosasaur.

NOTES

1. Jacques Derrida. *The Animal That Therefore I Am*. Ed. Marie-Louise Mallet, & Trans. David Wills (New York, NY: Fordham University Press, 2008), 16.
2. Ibid., 10.
3. Ibid., 16, my emphasis.
4. Ibid., 10.
5. Ibid., 18, my emphasis.
6. Ibid., 24.
7. Ibid., 150.
8. Joanna Hodge, *Derrida on Time* (London: Routledge, 2007), 185.

9. Derrida, *The Animal That Therefore I Am*, 24.

10. Ibid., 17.

11. Nicole Shukin, *Animal Capital: Rendering Life in Biopolitical Times* (Minneapolis, MN: University of Minnesota Press, 2009), 36.

12. Ibid., 37.

13. Ibid., 37.

14. Jacques Derrida, *The Beast and the Sovereign Vol. II*, Trans. Geoffrey Bennington (Chicago, IL: University of Chicago Press, 2011), 179.

15. Levi Bryant, Nick Srnicek and Graham Harman, "Towards a Speculative Philosophy," (In *The Speculative Turn: Continental Materialism and Realism*. Ed. Levi Bryant, Nick Srnicek and Graham Harman. Melbourne: re.press, 2011), 3.

16. Quentin Meillassoux, *After Finitude: An Essay on the Necessity of Contingency*, Trans. Ray Brassier (London: Bloomsbury, 2010), 26.

17. Timothy Morton, *Hyperobjects: Philosophy and Ecology After the End of the World* (Minneapolis, MN: University of Minnesota Press, 2014), 86.

18. Meillassoux, *After Finitude*, 7.

19. Ibid., 24, original emphasis.

20. Julius T. Csotonyi and Steve White, *The Paleoart of Julius Csotonyi: Dinosaurs, Saber-Tooths, and Beyond* (London: Titan Books, 2014), 47.

21. Kent A. Stevens, "Binocular Vision in Therapod Dinosaurs" (*Journal of Vertebrate Paleontology* 26.2, Jun. 12, 2006), 321.

22. Donald G. McNeil, Jr., "Where Art and Paleontology Intersect, Fossils Become Faces" (*The New York Times*, June 9, 2009), unpag.

23. To view Viktor Deak's paleoart, visit http://www.anatomicalorigins.com.

24. David Downs, "Neanderthal Sculptor Do Job Good, Smash Conventions" (*Gelf Magazine*, July14, 2009), unpag.

25. Jessica Mordsley, "Tracing Origins in Paleoanthropology" (*Oxford Literary Review* 29, 2007), 86.

26. Ibid., 88.

27. Ibid., 88, my emphasis.

28. Ibid., 95.

29. Eileen Joy, "You are Here: A Manifesto," In *Animal, Vegetable, Mineral: Ethics and Objects*, Ed. Jeffrey Jerome Cohen (New York, NY: Punctum Books, 2012), 166.

30. Serenella Iovino and Serpil Opperman, "Introduction: Stories Come to Matter." In *Material Ecocriticism*, ed. Serenella Iovino and Serpil Opperman (Bloomington, IN: Indiana University Press, 2014), 1.

31. Karen Barad, *Meeting the Universe Halfway: Quantum Physics and the Entanglement of Matter and Meaning* (Durham, NC: Duke University Press, 2007), 184.

32. Martin Rudwick, *The Meaning of Fossils: Episodes in the History of Paleontology* (Chicago, IL: University of Chicago Press, 1985), 1.

33. Donald Grayson, *The Establishment of Human Antiquity* (New York, NY: Academic Press, 1983), 5.

34. Jane Davidson, *A History of Paleontology Illustration* (Bloomington, IN: Indiana University Press, 2008), 51.

35. Ibid., 155.

36. Ibid., 183.

37. Richard Milner, *Charles R. Knight: The Artist Who Saw Through Time* (New York, NY: Abrams, 2012), 13.

38. Stephen Jay Gould, "Dinosaur Deconstruction" (*Discover Magazine*, October 1, 1993), unpag.

39. Ibid.

40. Ibid.

41. Ibid.

42. Michel Foucault, *The Order of Things: An Archaeology of the Human Sciences* (London: Routledge, 2002), 164–5.

43. Gould, "Dinosaur Deconstruction," unpag.

44. Jeremiah Dyehouse, "A Textbook Case Revisited: Visual Rhetoric and Series Patterning in the American Museum of Natural History's Horse Evolution Displays" (*Technical Communication Quarterly* 20.3, 2011), 327.

45. Samuel J. M. M. Alberti, "Maharajah the Elephant's Journey: From Nature to Culture" (In *The Afterlives of Animals: A Museum Menagerie*, Ed. Samuel J. M. M. Alberti. Charlottesville, VA: University of Virginia Press, 2011), 37.

46. Michel Foucault, *The Order of Things: An Archaeology of the Human Sciences*, 170.

47. Julius T. Csotonyi and Steve White, *The Paleoart of Julius Csotonyi: Dinosaurs, Saber-Tooths, and Beyond*, 128.

48. Ibid., 128.

49. Ibid., 128.

50. Dave Hone, qtd. in Julius Csotonyi and Steve White, *The Paleoart of Julius Csotonyi: Dinosaurs, Saber-Tooths, and Beyond*, 138.

51. Julius T. Csotonyi and Steve White, *The Paleoart of Julius Csotonyi: Dinosaurs, Saber-Tooths, and Beyond*, 136.

52. Timothy Morton, *Ecology without Nature: Rethinking Environmental Aesthetics* (Cambridge, MA: Harvard University Press, 2009), 16.

WORKS CITED

Alberti, Samuel J. M. M. "Maharajah the Elephant's Journey: From Nature to Culture." In *The Afterlives of Animals: A Museum Menagerie*. Ed. Samuel J. M. M. Alberti. Charlottesville, VA: University of Virginia Press, 2011.

Barad, Karen. *Meeting the Universe Halfway: Quantum Physics and the Entanglement of Matter and Meaning*. Durham, NC: Duke University Press, 2007.

Bryant, Levi, Srnicek, Nick, and Harman, Graham. "Towards a Speculative Philosophy," In *The Speculative Turn: Continental Materialism and Realism*. Ed. Levi Bryant, Nick Srnicek and Graham Harman. Melbourne: re.press, 2011.

Csotonyi, Julius T. and White, Steve. *The Paleoart of Julius Csotonyi: Dinosaurs, Saber-Tooths, and Beyond*. London: Titan Books, 2014.

Davidson, Jane. *A History of Paleontology Illustration*. Bloomington, IN: Indiana University Press, 2008.

Derrida, Jacques. *The Animal That Therefore I Am*. Ed. Marie-Louise Mallet, & Trans. David Wills. New York, NY: Fordham University Press, 2008.

———. *The Beast and the Sovereign Vol. II*. Trans. Geoffrey Bennington. Chicago, IL: University of Chicago Press, 2011.

Downs, David. "Neanderthal Sculptor Do Job Good, Smash Conventions," *Gelf Magazine*, July14, 2009. Accessed April 10, 2017. http://www.gelfmagazine.com/archives/neanderthal_sculptor_do_job_good_smash_conventions.php.

Dyehouse, Jeremiah. "A Textbook Case Revisited: Visual Rhetoric and Series Patterning in the American Museum of Natural History's Horse Evolution Displays."*Technical Communication Quarterly* 20.3 (2011). Accessed April 10, 2017.http://www.tandfonline.com/doi/abs/10.1080/10572252.2011.578235, 327.

Foucault, Michel. *The Order of Things: An Archaeology of the Human Sciences*. London: Routledge, 2002.

Gould, Stephen Jay. "Dinosaur Deconstruction."*Discover Magazine*. October 1, 1993. Accessed April 10, 2017. http://discovermagazine.com/1993/oct/dinosaurdeconstr284/.

Grayson, Donald. *The Establishment of Human Antiquity*. New York, NY: Academic Press, 1983.

Hodge, Joanna. *Derrida on Time*. London: Routledge, 2007.

Iovino, Serenella and Opperman, Serpil. "Introduction: Stories Come to Matter." In *Material Ecocriticism*, ed. Serenella Iovino and Serpil Opperman. Bloomington, IN: Indiana University Press, 2014.

Joy, Eileen. "You are Here: A Manifesto," In *Animal, Vegetable, Mineral: Ethics and Objects*. Ed. Jeffrey Jerome Cohen. New York, NY: Punctum Books, 2012.

McNeil, Donald G. Jr. "Where Art and Paleontology Intersect, Fossils Become Faces." *The New York Times*, June 9, 2009. Accessed April 10, 2017. http://www.nytimes.com/2009/06/02/science/02prof.html.

Meillassoux, Quentin. *After Finitude: An Essay on the Necessity of Contingency*. Trans. Ray Brassier. London: Bloomsbury, 2010.

Milner, Richard. *Charles R. Knight: The Artist Who Saw Through Time*. New York, NY: Abrams, 2012.

Mordsley, Jessica. "Tracing Origins in Paleoanthropology," *Oxford Literary Review* 29 (2007): 86–95. Accessed April 10, 2017. http://www.jstor.org/stable/44031016?seq=1#page_scan_tab_contents.

Morton, Timothy. *Hyperobjects: Philosophy and Ecology After the End of the World*. Minneapolis, MN: University of Minnesota Press, 2014.

———. *Ecology Without Nature: Rethinking Environmental Aesthetics*. Cambridge, MA: Harvard University Press, 2009.

Rudwick, Martin. *The Meaning of Fossils: Episodes in the History of Paleontology*. Chicago, IL: University of Chicago Press, 1985.

Shukin, Nicole. *Animal Capital: Rendering Life in Biopolitical Times*. Minneapolis, MN: University of Minnesota Press, 2009.

Stevens, Kent A. "Binocular Vision in Therapod Dinosaurs," *Journal of Vertebrate Paleontology* 26.2 (Jun. 12, 2006). Accessed April 10, 2017. https://www.jstor.org/stable/4524572?seq=1#page_scan_tab_contents.

Chapter 5

"The Dignity of Mankind"

Edward Tyson's Anatomy of a Pygmie *and the Ape-Man Boundary*

Nicole Mennell

In the "Eleventh Session" of *The Beast and the Sovereign: Volume I*, Jacques Derrida focuses on the autopsy of an elephant conducted under the sovereign gaze of King Louis XIV in 1681. He conjures the scene of the "enormous, heavy, poor beast, dead or killed [...] among doctors, surgeons, or other armed butchers, impatient to show what they could do but just as impatient to see and give to be seen what they were going to see, trembling with lust for autopsy."[1] The subjection of the elephant's body to this "autopsic" gaze emphasizes not only the sovereign power of a king but also the assumed sovereignty of humans over animals.[2] For humans to gain knowledge, the animal is reduced to an object of scientific research, making animal autopsies an enactment of human power. Ultimately, Derrida asserts, "[k]nowledge is sovereign" because it suggests "possession and mastery of its object."[3] However, as Kelly Oliver observes in her analysis of Derrida's "Eleventh Session," the animal looks back at the living because "the secrets of life itself, do not yield to the autopsic gaze. Rather, there is always something beyond the limits of human seeing, something exorbitant, beyond the limits of human understanding."[4] The inability to gain complete and certain knowledge of the nature of life and existence through the ocular apparatus, prevents the animal, subjected to the autopsic gaze, from being fully possessed and mastered by the human.

The elephant autopsy at the center of the "Eleventh Session" occurs at a particular moment in the history of scientific enquiry and human-animal relations; a moment characterized by an exceptional drive to understand and hierarchically systematize nature through anatomical studies of numerous life forms, many of which were newly discovered.[5] The "Eleventh Session" can therefore be used to reassess how animals, especially those regarded as curiosities, were dissected, analyzed, and categorized by anatomists in the early modern period in order to gain knowledge. Taking this into consideration, the present chapter will place Derrida's discussion of the autopsic gaze alongside that of the work undertaken by Edward Tyson, a physician, fellow of the Royal Society of London, and a pioneer of early modern comparative anatomy, who took a particular interest in rare animals. The focus of this chapter will be on perhaps his most significant text, *Orang-Outang, sive Homo Sylvestris: Or, the Anatomy of a Pygmie Compared with that of a Monkey, an Ape, and a Man.*[6]

Published in 1699, having been presented before the Royal Society in June 1698, Tyson's *Anatomy of a Pygmie* is the first comparative anatomy of the human body and what we now know was a chimpanzee. Since anthropoid apes had been largely unknown to the Western world until the mid-seventeenth century, it was unclear where they should be positioned in relation to humankind. The male juvenile chimpanzee's external likeness to the human caused Tyson to speculate whether it was the elusive pygmy, a hybridized creature which was said to resemble humans in ancient texts and contemporary traveler accounts. The pygmy's resemblance to the human extended to its musculoskeletal system and internal organs, leading to Tyson's conclusion that it "seems the Nexus of the Animal and Rational."[7] The anatomist therefore situated the pygmy in the gradational chain of being as an intermediary between humans and animals, making it the first text to provide an empirical foundation for the similarities between humans and primates, over a century before the publication of Darwin's *Origin of Species.*[8] While Tyson was aware that this categorization was remarkable (and wanted to be credited for having made this discovery), the pygmy's proximity was a source of anxiety for the anatomist because it troubled the human-animal boundary; a boundary built upon the Aristotelian-Scholastic conception of man as rational animal and the Christian belief in humans as being created in God's image.

In their respective articles on *The Anatomy of the Pygmie*, Erica Fudge and Susan Wiseman argue that Tyson's text is not an objective scientific study but one greatly influenced by the social and political concerns of the period in which it was produced.[9] Building upon their arguments, this chapter suggests that the autopsic gaze is employed by Tyson in an attempt to achieve mastery over an animal that, albeit deceased, looks back as a reminder of the vulnerability of human superiority in the natural order. Tyson's "lust for autopsy," his desire to see and know more about this curious animal, and by extension to gain greater knowledge of the human, motivates his study. However, his belief in human perfectibility obscures his vision, prompting a reliance on immaterial evidence when the pygmy becomes "too near akin."[10]

Taking its meaning from the Latin *autopsia*, as Derrida informs us, autopsy is defined as "the experience that consists in seeing with one's own eyes, and thus of being able to bear witness."[11]As seventeenth-century natural history was dominated by the ideals of Francis Bacon's New Science, which valued empirical evidence over information provided in former accounts, seeing the inner workings of animal bodies "with one's own eyes" gained greater importance. Connected to this new way of seeing animals was the rise of comparative anatomy in the mid-seventeenth century, of which Tyson was a key figure. Institutions, such as the Académie Royale des Science, whose members conducted the autopsy of Louis XIV's elephant, used comparative anatomy in order to, as Derrida recognizes, "clarify the analogies between human anatomy and physiology and animal physiology."[12] When considering the more symbolic analogies between Louis XIV and this sovereign beast, Derrida suggests, "One thinks of this elephenomenelephant that was no longer looking at them but that could have seen them, with its own eyes seen the king see it in its own autopsy."[13] What Derrida highlights in this reversal of the gaze, perhaps inadvertently, is the capacity of comparative anatomical studies to diminish the distance between the human and the animal, because the premise of comparative anatomy depends on a basic assumption of continuity between the two. Thus, while comparative anatomical studies demonstrate human mastery over the animal, the similarities they reveal paradoxically call into question the notion of human perfectibility.

To overcome the problem posed by these physical analogies, ideas of human uniqueness were maintained in the early modern period through the assertion that humans alone were endowed with rationality by God. Cartesian mechanical philosophy took this idea further, demarcating a complete separation of the visible body from the invisible rational soul. For example, in *Two Discourses Concerning the Soul of Brutes,* Thomas Willis asserts that the sensitive soul "which is Common to Brute Animals with Man [...] seems to depend altogether on the Body [...] and to be plainly Corporeal," whereas the rational soul is "Superior and Immaterial."[14] The reliance of human sovereignty upon the possession of an incorporeal, and therefore invisible, rational soul can be regarded as, to use Oliver's term, the "blind spot" of the autopsic gaze, which is "both necessary and unsettling to it."[15] The immateriality of the rational soul is essential because it allows humans to uphold the notion of their superiority.

In his role as a comparative anatomist, Tyson strongly advocated the empirical way of seeing and knowing gained through autopsy, and adhered to the principles of Cartesian mechanical philosophy. In *Phocæna, or The Anatomy of a Porpess*, published in 1680, Tyson criticizes "Former Naturalists" for "relying on others, when Autopsie, and their own Experience can only inform them; and their Conscience and eyes may be as a thousand Witnesses."[16] One year after *Phocæna* was published, Tyson translated Jan Swammerdam's *Ephemeri Vita, or, The Natural History and Anatomy of the Ephemeron, a Fly that Lives but Five Hours*. In the preface to this text, Tyson reinforces the necessity of individual anatomical studies in creating a new and more accurate natural history, claiming: "if we would understand how 'tis that Nature gives Life and Motion to these *Automata*, we must unloose the Case, and take asunder the several Wheels and Springs, and carefully observe how she joyns them all together."[17] In this evocation of Cartesian philosophy, Tyson identifies animals as objects to be meticulously studied in, to quote Derrida, "the service of knowledge."[18] Through his methodical examination of individual cases, Tyson sought to create "a scale of Gradation," beginning with zoophytes and ending with "Man and Intelligent Beings, that are a boundary between Divinity and the Creation," asserting that such studies would "enrich the mind with the knowledge of God in his works, and of our selves."[19] Tyson's anthropocentrically driven studies therefore fulfill the additional definition Derrida gives of *autopsia*: "that of participation in divine power."[20] As well as enacting man's ascendancy over nature, Tyson presents comparative anatomy as a method of moving humans closer to God by revealing an insight into the grand design of creation. Despite his declared adherence to the values of empiricism, Tyson's overriding ambition to demonstrate the privileged position of humans in the gradational chain of being creates a "blind spot" in his autopsic gaze.

Tyson's preoccupation with the unique status of the human can be linked to his appointment in 1684 as medical superintendent of Bethlem Hospital.[21] It is striking that, despite the shifts in the history of madness, people suffering with perceived disorders of mind or soul have been consistently dehumanized due to the supposed loss of their reason. Derrida draws attention to this in the "Eleventh Session" when he asserts that "curiosity" can be used to "formalize the whole field of analogy that gathers the beast and the madman"; both animals and those considered insane, Derrida states, "become, in all sense of the words, *curiosities* for the eager, compulsive *curiosity* of [...] those who are outside and approach them only to within a certain distance to observe or inspect them in a sovereign manner from outside after having locked them up."[22] What gives those observers their "sovereign manner," particularly in the early modern context, is their self-perceived possession of rationality. In relation to this, a preacher delivering a sermon in 1719 described the patients of Bethlem as having "lost their Senses, and the Use of the Reason and Understanding which they once had; and now had nothing left but their outward Shape to distinguish them from the Creatures below them."[23] Tyson is credited with restoring his patient's humanity through his innovative reorganization of Bethlem. This was achieved through, among other therapeutic initiatives, the establishment of a wardrobe fund, a facility for outpatients, and a relief fund for discharged patients, commonly known as "Tyson's Gift."[24] Furthermore, an order dated August 11, 1699, enforced a new regulation for attendants to, "take effectual Care that no Person whatsoever that they in the least suspect to be lewd or disorderly person, [...] and come there to idle away their Time, be upon any account permitted to come into the said Hospital."[25] Despite the continued admittance of the public (largely for financial reasons as those who came to view Bethlem's patients usually became its benefactors), the regulation represents a concerted effort to prevent those of the "lower sort" being allowed to view the patients as curiosities, thus protecting the human status of those under the care of Bethlem's physicians. Tyson's concern for his patients' wellbeing is made apparent in Elkanah Settle's *Threnodium Apollinare, A funeral poem to the memory of Dr. Edward Tyson*: "To sickning Minds did his Great ART extend. / [...] To Depraved *Sense* / Great *TYSON'S* Power new *Organs* cou'd dispense; / Lost *Reason* reinthrone."[26] Settle credits Tyson with reestablishing human sovereignty through the symbolic image of his patients' reason being "reinthrone[d]." There is a significant correlation between Tyson's endeavor to ensure the place of humans at the top of the natural order through his work at Bethlem and his treatise which seeks to reinforce the animality of the pygmy; two concerns which center on the rationality and dignity of humans.

Taking into account Tyson's work at Bethlem, and his ambition to construct a hierarchical chain of being that positioned humans as the sovereign species, the pygmy and its uncanny resemblance to mankind comes to represent more than just a curious scientific case, but carries implications for the assumed supremacy of humans. As Derrida outlines in the "Tenth Session" of *The Beast and the Sovereign* when discussing La Fontaine's fable of *The Monkey and the Dolphin*, some animals, specifically those considered to be anthropoid, are "more human than others" and therefore more problematic.[27] It is because of their anatomical, intellectual, and emotional similarities that apes have long held the gaze of humans. They have consequently been exhibited as curiosities to be gawked at in various forms of public entertainment and scientific enquiry, simultaneously acting as both a source of intrigue and unease for their closest living relatives. Not only do apes physically resemble humans, they also possess cognitive abilities, the possession of which has historically been viewed as the defining marker of humanity. Apes encroach arguably more so than other animals upon, what Derrida terms in *The Animal That Therefore I Am*, the "unilinear and indivisible line" that is perceived to exist between human and animal, exposing it as a "multiple and heterogeneous border."[28]

Apes also generated interest and unease in the early modern period due to their supposed existence on the boundary that separated human and animal.[29] That apes were viewed as abnormalities is demonstrated by their designation in medieval bestiaries as "monsters," a category to which Pliny also assigned pygmies.[30] This classification was largely due to the stories surrounding the origins of primates, which conform to Derrida's assessment of such fables and their "fabulization" of the animal as being "a discourse *of* man, on man, indeed on the animality of man, but for and in man."[31] For example, the Ovidian tale of the Cercopians' transformation into apes is imbued with ambivalence due to the unsettling resemblance apes bear to humans, which is captured most forcefully in the description, "Although unlike, appearing like to men."[32] Perhaps more importantly, the Cercopians are denied the ability to speak, the faculty that Aristotle linked to the possession of a soul in his claim: "[n]ot every sound made by an animal is voice (for it is possible to make a sound also with the tongue or as in coughing); but that which does the striking must have a soul and there must be a certain imagination (for voice is a particular sound which has meaning [...])."[33] Descartes furthered the Aristotelian position through his assertion that animals' lack of speech "shows that they have no reason at all."[34] Although apes were not believed to possess what Descartes considered true speech, along with other nonhuman animals which did not fit clearly into established categories, they were perceived as dangerous in the medieval and early modern period because they represented the possibility of human regression into animality.[35]

Similarly, the medieval debates which surrounded the existence of pygmies were concerned with their place in relation to both humankind and apes, because their anthropomorphic descriptions in ancient texts positioned them a degree closer to humans.[36] In his article that discusses the influence of these debates on Tyson's text, Hans Thijssen notes that Saint Augustine's discussions of whether Pliny's monstrous races actually existed prompted the more unsettling question of whether they were also Adam's descendants.[37] In addition, it was believed that not only did pygmies look like humans but that they, unlike apes, could speak. In his translation of Conrad Genser's *Hisoriae Animalium*, Edward Topsell discusses how "*Niphus* proueth that they are not men bycause they haue no perfect vse of reason, no modesty, no honesty, nor iustice of government, and although they speake yet is their language imperfect."[38] Niphus's description seems to define pygmies straightforwardly as apes, however, their alleged ability to speak, albeit "imperfect[ly]," problematizes this conclusion because it suggests they have cognitive abilities. It was due to the attribution of language to pygmies that the debates surrounding their status in relation to humankind were not easily concluded.

Writing his *Anatomy of a Pygmie* within a culture still very much influenced by mythical stories of pygmies and the medieval debates which surrounded their anthropomorphic qualities in ancient accounts, Tyson uses his dissection to prove that pygmies were mistaken interpretations of real productions of nature rather than complete fictions. He sought to achieve this through the inclusion of *A Philological Essay Concerning the Pygmies, the Cynocephali, the Satyrs, and the Sphinges of the Ancients. Wherein it will appear that all are either Apes or Monkeys, and not Men, as formerly pretended.* In relation to this, Tyson explains that he did not identify the animal as a pygmy in order to "divert the Reader, with the Recital of the Fabulous and Romantick Stories" and clarifies, "as my chief Design in this Understanding is the Improvement of the *Natural History of Animals*; so I have made it my Bussiness more, to find out the *Truth*, than to enlarge in the *Mythology*."[39] This disclaimer asserts that the treatise is not a continuation of false stories and justifies using the myth of the pygmy for a greater cause, the enlargement of natural history. Tyson therefore utilizes the curiosity in the pygmy and other mythical creatures to intrigue his colleagues and thereby further taxonomic studies of apes, using the stories found in popular literature as a call to arms. This objective is apparent in his inclusion of Andrew Battell's account of the Pongo and Engeco "monsters," which are seen to be "in all proportions like a man."[40] The anatomist concludes that while these creatures are "different from our Subject," he hopes they "might excite some *Inquisitive Observers* to give us a truer Account of this large and noble *Species* of *Animals*."[41] Although Tyson is concerned with policing the human-animal boundary, he also toys with the difficulties the pygmy presents in order to justify his overall aim, without fully collapsing the human-animal distinction.

A close reading of the *Philological Essay* Tyson attaches to his anatomical study reveals his invested interest in human sovereignty. He is especially critical of the poets and historians' use of the myths surrounding pygmies because of their potential animalization of the human, and suggests their stories originate in Homer's *Illiad,* "wherein he compares the Shouts that the *Trojans* made, when they were going to joyn the Battle with the *Græcians*, to the great Noise of the Cranes, going to fight the *Pygmies*."[42] Tyson accepts Homer's use of this simile "for the Beauty of his *Poem*" and believes that others have mistaken this to mean that pygmies were actually men.[43] There is a noticeable shift in tone from Tyson's discussion of Homer to his description of "the fabulous *Historians* in after Ages, who have so oddly drest up this Story by their fantastical Inventions," and the "*Moderns* [...] that have maintained that these *Pygmies* were real men."[44] One such example is Joshua Barnes's *Gerania: A New Discovery of a Little sort of People Anciently Discoursed of, called Pygmies,* published in 1675. In the preface to his text, Barnes refers to Pliny's attribution of such wonders to the heat of the climate in which they were produced, claiming that "every Climate doth peculiarly afford something rare and unusual, it stands with reason, that some Climate should in especial manner produce some more extraordinary Novelty."[45] Tyson opposes this belief with his resolute assertion that it is "not the *heat* of the Country, but the warm and fertile Imagination of these *Historians*, that has been productive of them."[46] Tyson's anger is directly linked to the misapplication of the title "Man" to pygmies, causing him to declare that these writers are "so wretchedly whimsical, in abusing the Dignity of Mankind, by giving the Name of *Man* to such monstrous Productions of their idle imaginations."[47] By claiming humanity for these creatures the poets and historians threaten human uniqueness. In this instance, it is not the pygmy itself which is "monstrous," but the stories that the authors have created.

Tyson also disparages travelers and merchants for abusing the credulity of the public by falsely representing apes as "the *Men* of those Countries from whence they came."[48] Notably, the animal at the centre of Tyson's study, which was brought from Angola to England in 1698, dying shortly after its arrival, was possibly displayed as a human-animal hybrid at Moncress' Coffee House in June 1698. Richard Nash argues that the pygmy is referred to in a handbill as, "a MONSTER that lately died there, being Humane upwards, and Bruit downwards."[49] In the "Eleventh Session," Derrida is interested in those animals which, "captured and transported, became curiosities for the curiosity of the curious who wanted and henceforth were able to see and observe them with their own eyes, autoptically."[50] This suggests the indulgence of curiosity through observing the external particulars of animal bodies, a way of seeing animals which is markedly different from the penetrating nature of the autopsic gaze. To further this point, it is important to note that the term "curiosity" had specific associations in early modern science. In her article on this topic, Lorraine Daston convincingly argues that from the mid-sixteenth century onwards curiosity was transformed from a vice, specifically a variety of lust, into a virtue, through its associations with the self-disciplined, but still "insatiable," drive for knowledge. Daston suggests that this particular form of curiosity was distinctive in its application to seventeenth-century investigations of nature as it positioned "the natural philosopher as lynx-eyed observer and as opponent to the vulgar."[51] If we accept the monster displayed at Moncress' Coffee House as Tyson's chimpanzee, he counters its commercially driven and false representation through his anatomical study, which is given authority through its associations with the Royal Society, a sovereign institution of knowledge and regulated curiosity.[52] By framing the autopsy of the pygmy as, to use Derrida's words, "a learned objectification in the academic service of a learned society," Tyson transforms the animal from a monstrous curiosity to be gazed upon by the masses into a curious specimen to be studied by men of science.[53]

In undermining the representation of the pygmy as a hybrid form of humanity, Tyson is able to more securely present the analogies between the pygmy's anatomy and that of the human body as scientifically curious. He concludes his study with a list of two catalogues, one that shows in forty-eight points that his "Pygmie *more resembled a* Man, *than* Apes *and* Monkeys do," and one that details the thirty-four points it "*diffr'd from a* Man."[54] One of the most noteworthy similarities Tyson discovered in his comparative anatomy, albeit inaccurately, is that in the pygmy "the *Peritonæum* [. . .] [is] formed as 'tis in *Man*; [. . .] as if *Nature* did design it to go *erect*."[55] This is a significant observation as the human bipedal stature is used to indicate their superiority over all other beings. Justin Smith notes the importance of human bipedalism due to the perceived correlation of "the capacity for higher cognition with the ability to walk on two feet."[56] Although a bipedal animal calls this correlation into question, Tyson is unconcerned by the animal's perceived ability to walk erect because this makes the error of the ancients in attributing human status to pygmies (always depicted as bipeds in the myths), more plausible.

The images Tyson includes of the pygmy as a biped, created by the Flemish engraver Michael van der Gucht, also act as evidence for how easily such an animal could be mistaken for a human (see Figure 5.1). Alex Potts argues that, while early modern depictions of animals in natural history were used

Figure 5.1 Michael van der Gucht, "Figure the First," unpaginated plate in Tyson's *Orang-Outang, sive Homo Sylvestris: Or the Anatomy of a Pygmie*, 1699. Engraving. British Library. © 2010 The British Library Board.

to provide accurate observations, "[n]o images, even scientific illustrations, can be purely neutral records."[57] The same assessment can be applied to the images of the pygmy, which are included to satisfy the curiosity of the learned observer by highlighting the animal's anatomical similarities, providing additional evidence for its categorization as the "Nexus of the Animal and the Rational." The first figure represents an animate being which looks back at the viewer with a direct gaze and the subtle hint of a smile. However, Tyson controls the interpretation of this image through the accompanying descriptions he provides; for example, in the statement, "The *Face* looks like an Old wither'd Man's," the preposition "like" maintains a degree of difference, reinforcing the animal's categorization beneath the human.[58] Tyson also explains that "[b]eing weak, the better to support him, I have given him a stick in his Right-Hand."[59] There is more to the allocation of a walking stick than Tyson's justification suggests as this was a traditional motif used in depictions of apes.[60] An instance of this is the image of a "satyr, simian or wild man," found in Conrad Gesner's *Icones Animalium*, which Tyson reproduces in his text.[61] The inclusion of the walking stick therefore implicitly identifies the pygmy as an ape. Furthermore, the first and second figures are preceded by a reproduction of the animal's anatomical deconstruction, beginning with anterior and posterior depictions of its musculature, the animal is stripped down to its skeleton, and finally to its separate organs. Although the first image provides an opportunity for the gaze to be reversed, the subsequent images in their visual dismantling of the pygmy's body reduce it, like Louis XIV's elephant, to a "cadaver docile to autopsy," thus reaffirming Tyson's mastery of the animal.[62]

The reflection of the human in the animal becomes more problematic when Tyson turns to the pygmy's interior anatomical similarities, specifically that of its brain and organs of speech. In Derrida's discussion of Descartes' writings on the animal, which inform the "Eleventh Session," he suggests that "[w]hat is subject to critique [. . .] is the passage from outside to inside, belief in the possibility of inducing from this *exterior* resemblance an *interior* analogy, namely, the presence in the animal of a soul, of passions like our own."[63] Conversely, Tyson does reflect on the pygmy's passions through observations of its expression of "Joy and Grief," stating "I heard it *Cry* my self like a *Child*; and he hath been often seen to kick with his Feet, as Children do, when either he was pleased or angered."[64] Despite the comparison made between the pygmy and the manner in which children communicate their feelings, Tyson does not entertain any suggestion that the animal has the ability to speak. Acknowledging that the pygmy possesses the appropriate vocal organs, Tyson claims, "if there was any farther advantage for the forming of *Speech*, I can't but think our *Pygmie* had it. But upon the best Enquiry, I was never informed, that it attempted any thing that way."[65] The admittance of a voice to the pygmy would suggest an "*interior* analogy" due to the early modern belief, to quote Tyson, "*that Speech is an Action more peculiar to Man, and which most distinguished him from Brutes*."[66] It therefore becomes vital for Tyson to preserve speech as a distinctive feature of human reason.

As it was believed the brain was the "Seat of the *Soul* it self," it is unsurprising that Tyson applies the same logic regarding speech to the similarities between the human brain and that of his specimen. He asserts that, "since there is so great a disparity between the *Soul* of a *Man*, and a *Brute*, the *Organ* likewise in which 'tis placed should be very different too."[67] The pygmy's anatomy has proved otherwise, causing Tyson to declare: "those *Nobler Faculties* in the *Mind* of *Man*, must certainly have a *higher Principle* [...] for why else, where the *Organ* is the same, should not the *Actions* be the same too? and if all depended on the *Organ*, not only our *Pygmie*, but other *Brutes* likewise, would be too near akin to us."[68] The visual similarities revealed by his anatomical study cause Tyson to depart from his reliance upon the empirical evidence and turn to Cartesian dualism, in order to claim a separation of the material body from the immaterial soul. In doing so, Tyson polices the boundary between the human and all animals, as he is aware of the consequences that his study would have for the unique status of the human.

Tyson also uses anecdotes about the animal's behavior when alive in an attempt to maintain the human-animal divide. Fudge rightly argues that because of these anecdotes "[t]here is a constant, and jarring, anthropomorphism at work which Tyson both uses and denies."[69] To begin with, describing the pygmy as "the most gentle and loving Creature that could be," Tyson goes on to state, "I was informed, tho' there were *Monkeys* aboard, yet 'twas observed he would never associate with them, as if nothing a-kin to them, would always avoid their Company."[70] Here the pygmy is referred to in the personal pronoun, "he," and is endowed with a sense of agency as it is seen to be tender and sociable with humans but refuses to associate with monkeys, as though the chimpanzee was aware of the position Tyson had given it above them in the hierarchical chain of being. However, Tyson denies the closeness of this animal to the human through a further anecdote in which the pygmy, like the monkey in La Fontaine's fable, is seen to imitate man inadequately. Tyson describes how "It would lie in a Bed, place his Head on the Pillow, and pull the Cloaths over him, as a Man would do; but was so careless, and so very a Brute, as to do all Nature's Occassions there."[71] Tyson appears to be aware of the anthropomorphism the anecdotes exhibit and, as Fudge notes, counters this with an account of the animal, referred to in the impersonal pronoun "it," soiling itself in order to reinforce its bestial nature.[72] Tyson's observations of the animal's behavior is redolent of Derrida's assessment of Cartesian philosophy as viewing the animal as an "object for a human who says 'I am,' spectacle for a specular subject who reflects his essence and who does not find, or does not want to find himself reflected in the image of the animal that he looks at but that doesn't look at him."[73] Tyson does find the essence of the human reflected in the pygmy, prompting the careful framing of these anecdotes in order to deny "the presence in the animal of a soul."

It is important to note that towards the end of 1701, Tyson had the opportunity to observe another primate, a mandrill which he refers to as a "Mantoger."[74] Despite its numerous similarities to the *Anatomy of a Pygmie*, Tyson's encounter with the mandrill, preserved in manuscript in the Tyson Folio at the Royal College of Physicians, has received little critical attention as it was not published in its entirety, with only the animal's physical description being reproduced in "The Anatomy of a Male Opossum."[75] Identifying this animal as a rarity, Tyson speculates whether it is the mythical manticore, which also troubled the human-animal boundary as it was said to resemble a man but to be of "the buigness of a Lyon" and to have "a Tayle like an Epirobian Scorpion."[76] The mandrill therefore provides Tyson with another opportunity to attack the poets and historians for "so disfigur[ing] the Truth, y^{t} there is no knowing it till one has pulled of ye Mask."[77] However, Tyson's "lust for autopsy" is denied satisfaction and he is only able to satisfy his curiosity by gazing upon the mandrill's external appearance and behavior.

In this account, Tyson presents the mandrill as being worthy of further study because it risks being perceived as a hybridized form of humanity. He observes that "its Eyes were very large near as big as a Man [...] & if any Stranger Looked intensely upon him, he would seem to be moved to make a Noise & nock his Stick on ye ground at him."[78] Comparable to his treatment of the pygmy, Tyson denies the mandrill's closeness to the human by emphasizing its physical differences, which is mainly achieved through the detailed pencil drawing contained in the manuscript (see Figure 5.2).[79] Despite also leveling a direct gaze at the viewer, the image of the mandrill sitting on its hindquarters represents an animal which does not visually undermine human uniqueness to the same extent as the anthropomorphized image of the bipedal pygmy. Tyson is keen to ensure the image is seen in this way by drawing attention to the ocular differences it depicts. Firstly, he describes the "remarkable [...] Colour" of the mandrill's nose as "so fierce a red" and the "azure or purple Colour" of its "face or Cheek." More significantly, he suggests that the animal is not a biped, claiming that "When it stands erect (which it cant do of its self but is supported by holding a Stick in its hands) tis crawling and week." Tyson therefore concludes that "its com[m]on posture is sitting upon its buttocks," as is illustrated in the accompanying image. He also emphasizes the mandrill's animality through the description of its "fierce" and untameable nature, remarking that the animal had "lamed its keeper in his hand & bitt him more than once. So y^{t} he was forced to put on a Muzzle

Figure 5.2 ***Mantichora,* c. 1701.** Pencil drawing. Royal College of Physicians, MS 618. © Royal College of Physicians.

on his Mouth." Comparable to the *Anatomy of a Pygmie*, Tyson's account of the mandrill's appearance and behavior is not a neutral record as it is used to discredit the myth of the manticore and, by extension, reestablish the human-animal boundary.

Tyson's encounter with the mandrill highlights his continued interest in animals that challenge the idea of human perfectibility. While he is able to firmly designate the mandrill as being of "ye Classis of the ye Apes," the pygmy is more problematic. Tyson states that although the pygmy "in the formation of the Body, and in the *Sensitive* or *Brutal* Soul" more resembles "a Man, than any other *Animal*," he takes "him to be wholly a *Brute*."[80] Despite this seemingly resolute conclusion, the pygmy's remarkable resemblance to the human, both externally and internally, does not allow for such a stable categorization. The notion of human sovereignty is only upheld by Tyson relying upon unobservable evidence and turning a blind eye to the more unsettling implications of the pygmy's proximity to the human.

Edward Tyson's *Anatomy of a Pygmie* is not solely concerned with proving his specimen is an animal but also with protecting the "Dignity of Mankind." By placing the pygmy under the lens of the autopsic gaze, Tyson reduces the animal to an object to be dissected and analyzed, reinforcing its position beneath humans in the gradational chain of being. The anatomist argues that although it "has many Advantages above the rest of [its] *Species*, yet I still think it but a sort of *Ape* and a meer *Brute* and as the *Proverb* has it [. . .] *an Ape is an Ape, tho' finely clad*."[81] The pygmy is put back in the place to which Tyson believes it belongs, as a passive and ridiculous imitator of humanity. In contrast, he declares that "*Man* is part *Brute*, part an *Angel*," positioning humans at the top of the natural order, the closest being to the divine.[82] Tyson's investment in the distinctive status of the human can be linked to his role at Bethlem Hospital and his endeavor to reconfirm the patients' humanity by restoring their senses and preventing them from being a source of degrading entertainment. Tyson's efforts as a physician parallel his anatomical study of the pygmy, through which he transforms the animal from a wondrous spectacle into a naturalized scientific specimen to be observed by the curious but learned eye. Nonetheless, by claiming that the pygmy is the "Nexus of the Animal and Rational," he fails to fully render the pygmy, to use Derrida's term, a "docile" object. Tyson finds the human reflected in the pygmy, and the unease this closeness creates is apparent in his need to emphasize human sovereignty.[83]

NOTES

1. Jacques Derrida, *The Beast and the Sovereign*: *Volume I*, ed. Michel Lisse, Marie-Louise Mallet, and Ginette Michaud, and trans. Geoffrey Bennington (Chicago, IL: University of Chicago Press, 2009), 284.

2. Derrida, *Beast*, 296.

3. Ibid., 280.

4. Kelly Oliver, "Elephant Eulogy: The Exorbitant Orb of an Elephant." In *The Animal Question in Deconstruction*, ed. Lyn Turner (Edinburgh: Edinburgh University Press, 2013), 91.

5. See Andrew Cunningham, "Animal Bodies and Comparative Anatomy," in *The Anatomist Anatomis'd*: *An Experimental Discipline in Enlightenment Europe* (Farnham: Ashgate, 2010), 195–359.

6. Edward Tyson, *Orang-Outang, sive Homo Sylvestris: Or, the Anatomy of a Pygmie Compared with that of a Monkey, an Ape, and a Man. To which is added, A Philological Essay Concerning the Pygmies, the Cynocephali, the Satyrs, and the Sphinges of the Ancients. Wherein it will appear that all are either Apes or Monkeys, and not Men, as formerly pretended* (London: Thomas Bennett and Daniel Brown, 1699). As the two sections have different signatures, the respective parts will be referred to as '*Anatomy*' and '*Philological*' in all further references.

7. Tyson, "The Epistle Dedicatory," in *Anatomy*, page unsigned.

8. F. J. Cole, *A History of Comparative Anatomy: From Aristotle to the Eighteenth Century* (London: Macmillan, 1944), 219; Charles Darwin, *The Origin of Species*, ed. Gillian Beer (Oxford: Oxford University Press, 1996).

9. Erica Fudge, "Bad Manners at the Anatomist's Table: Tyson's Pygmie and the Naturalisation of Truth," in *The Political Subject: Essays on the Self from Art, Politics and Science*, ed. Wendy Wheeler (London: Lawrence and Wishart Ltd, 2000), 13–40; Susan Wiseman, "Monstrous Perfectibility: Ape-Human Transformations in Hobbes, Bulwer, Tyson," in *At the Borders of the Human: Beasts, Bodies and Natural Philosophy in the Early Modern Period*, ed. Erica Fudge, Ruth Gilbert and Susan Wiseman (Basingstoke: Macmillan Press, 1999), 215–38.

10. Tyson, *Anatomy*, sig.[H4^{r}].

11. Derrida, *Beast*, 293.

12. Ibid., 283.

13. Ibid., 282.

14. Thomas Willis, *Two Discourses Concerning the Soul of Brutes, which is that of the Vital and Sensitive of Man* (London: Thomas Dring, 1683), sig.B^{r}.

15. Oliver, "Elephant Eulogy," 90.

16. Edward Tyson, *Phocæna, or The Anatomy of a Porpess Dissected at Gresham College: with a Præliminary Discourse concerning Anatomy, and a Natural History of Animals* (London: Benjamin Tooke, 1680), sig.[B3^{v}]. For further discussion, see Noah Moxham, "Edward Tyson's *Phocaena*: A Case Study in the Institutional Context of Scientific Publishing," *Notes and Records of the Royal Society* 66, no. 3 (2012): 235–52.

17. Edward Tyson, "To the Reader," in *Ephemeri Vita, or, The Natural History and Anatomy of the Ephemeron, a Fly that Lives but Five Hours*, by Jan Swammerdam (London: Henry Faithorne and John Kersey, 1681), sig.A2^{v}.

18. Derrida, *Beast*, 287.

19. Tyson, *Phocæna*, sig.C2^{r}, sig.C2^{v}.

20. Derrida, *Beast*, 296.

21. For further discussion, see Bad Manners, 21–25.

22. Derrida, *Beast*, 297, 297.

23. John Stow, *A Survey of the Cities of London and Westminster*, vol. 1 (London: A. Churchill, 1720), 197.

24. Jonathan Andrews *et al.*, *The History of Bethlem* (London: Routledge, 1997), 270–71.

25. Stow, *Survey*, 193. See also Ashley Montagu, 'Introduction," in *Orang-Outang, sive Homo Sylvestris: or the Anatomy of a Pygmie, 1699: A Facsimile with an Introduction by Ashley Montagu* (London: Dawsons of Pall Mall, 1966), 7.

26. Elkanah Settle, *Threnodium Apollinare. A Funeral Poem to the Memory of Dr. Edward Tyson* (London: Thomas Bickerton, 1708), sig.C^{r}-[C^{v}].

27. Derrida, *Beast*, 254.

28. Jacques Derrida, *The Animal That Therefore I Am*, ed. Marie-Louise Mallet, and trans. David Wills (New York, NY: Fordham University Press, 2008), 31.

29. For further discussion, see H. W. Janson, *Apes and Ape Lore in the Middle Ages and the Renaissance* (London: The Warburg Institute, 1952).

30. Wiseman, Monstrous Perfectibility, 217.

31. Derrida, *Animal*, 37.

32. Ovid, *Ovid's Metamorphosis Englished, Mytholigized and Represented in Figures by George Sandys* (Oxford: John Lichfield, 1632), l.118–25.

33. Aristole, *Aristotle's De Anima: Books II and III,* trans. D. W. Hamlyn (Oxford: Clarendon Press, 1968), II.8, 420^{b}27.

34. René Descartes, *Discourse on the Method and Meditations*, trans. Elizabeth S. Haldane and G. R. T. Ross (New York: Dover Publications, 2003), 39.

35. See Joyce E. Salisbury, *The Beast Within: Animals in the Middle Ages*, 2nd ed. (London: Routledge, 2011), 123.

36. J. M. M. H. Thijssen, "Reforging the Great Chain of Being: The Medieval Discussion of the Human Status of 'Pygmies' and its Influence on Edward Tyson," in *Ape, Man, Apeman: Changing Views since 1600,* ed. Corbey Raymond and Bert Theunissen (Leiden: Leiden University, 1995), 43.

37. Thijssen, "Reforging," 44.

38. Edward Topsell, *The Historie of Foure-Footed Beastes* (London: William Jaggard, 1607), sig.B2^{r}.

39. Tyson, *Anatomy*, sig.A^{r}.

40. Andrew Battell, *The Strange Adventures of Andrew Battell, of Leigh, in Angola and the Adjoining Regions. Reprinted from Purchas his Pilgrimes*, ed. E.G. Ravenstein (London: Hakluyt Society, 1901), 54.

41. Tyson, *Anatomy,* sig.[D3^{v}].

42. Tyson, *Philological*, sig.[A4^{v}].

43. Tyson, Ibid., sig.B^{r}.

44. Tyson, Ibid., sig.[B2^{v}], D^{r}.

45. Joshua Barnes, *Gerania: A New Discovery of a Little sort of People Anciently Discoursed of, called Pygmies, With a Lively Description of Their Stature, Habit, Manners, Buildings, Knowledge, and Government, being very delightful and profitable* (London: W.G for Obadiah Blagrave, 1675), sig.A3^{r}-[A3^{v}].

46. Tyson, *Philological*, sig.[B3^{r}].

47. Tyson, Ibid., sig.[C4^{v}].

48. Tyson, Ibid., sig.[F^{v}].

49. Handbill, "Advertisement, June, 1968, At Moncreff's Coffee house," 2026.25 (9), British Library. Quoted in Richard Nash, "Tyson's Pygmie: The Orang-outang and Augustan 'Satyr'," in *Ape, Man, Apeman: Changing Views since 1600,* ed. Corbey Raymond and Bert Theunissen (Leiden: Leiden University, 1995), 55.

50. Derrida, *Beast*, 298.

51. Lorraine Daston, "Curiosity in Early Modern Science," *Word and Image* 11, no. 4 (1995): 395, 392.

52. For further discussion, see Nash, Tyson's Pygmie, 55.

53. Derrida, *Beast*, 296.

54. Tyson, *Anatomy,* sig.[N2^{v}-N4^{r}].

55. Ibid., sig.[E^{v}]-E2^{r}.

56. Justin E. H. Smith, "Language, Bipedalism and the Mind–Body Problem in Edward Tyson's *Orang-Outang* (1699)," *Intellectual History Review* 17, no. 3 (2007): 291.

57. Alex Potts, "Natural Order and the Call of the Wild: The Politics of Animal Picturing," *Oxford Art Journal* 13, no. 1 (1990): 12.
58. Tyson, *Anatomy*, sig.[C4^{v}].
59. Ibid., sig.[C4^{v}].
60. See Robert Wokler, "Tyson and Buffon on the Ourang-Outang," *Studies on Voltaire and the Eighteenth Century* 155, no. 5 (1976), 2302.
61. Conrad Gesner, *Icones Animalium* (Zurich: Christof Froshover, 1560), 95. See fig. 17 in *Anatomy*, page unsigned.
62. Derrida, *Beast*, 282.
63. Derrida, *Animal*, 79.
64. Tyson, *Anatomy*, sig.E^{r}.
65. Ibid., sig.H2^{r}.
66. Ibid., sig.H2^{r}.
67. Ibid., sig. [H4^{r}].
68. Ibid., sig.[H4^{r}].
69. Fudge, 19.
70. Tyson, *Anatomy*, sig.[B4^{r}].
71. Ibid., sig.[B4^{v}].
72. See Fudge Bad Manners, 26.
73. Derrida, *Animal*, 82.
74. Tyson Folio, MS 618, Royal College of Physicians, fol. 86.
75. William Cowper and Edward Tyson, "Carigueya, Seu Marsupiale Americanum Masculum. Or, The Anatomy of a Male Opossum," *Philosophical Transactions* 24 (1704–1705), 1571–72.
76. MS 618, fol. 32.
77. MS 618, fol. 32.
78. MS 618, fol. 86. All further references are to fol. 86.
79. For further discussion of this image, see Ashley Montagu, *Edward Tyson, M. F., M.D., F.R.S. 1650-1708, and the Rise of Human and Comparative Anatomy in England: A Study in the History of Science* (Philadelphia: The American Philosophical Society, 1943), 329.
80. Tyson, *Anatomy*, sig.B2^{v}.
81. Ibid., sig.[M^{v}].
82. Ibid., sig.[H4^{r}].
83. I am grateful to Shanyn Altman, Emily Bartlett, Camilla Bostock, Andrew Hadfield, and Noah Moxham for reading earlier drafts of this chapter.

WORKS CITED

Andrews, Jonathan, Asa Brigs, Roy Porter, Penny Tucker, and Keir Waddington. *The History of Bethlem*. London: Routledge, 1997.

Aristole. *Aristotle's De Anima: Books II and III*, translated by D. W. Hamlyn. Oxford: Clarendon Press, 1968.

Barnes, Joshua. *Gerania: A New Discovery of a Little Sort of People Anciently Discoursed of, called Pygmies, With a Lively Description of Their Stature, Habit, Manners, Buildings, Knowledge, and Government, Being Very Delightful and Profitable.* London: W.G for Obadiah Blagrave, 1675.

Battell, Andrew. *The Strange Adventures of Andrew Battell, of Leigh, in Angola and the Adjoining Regions. Reprinted from Purchas his Pilgrimes*, edited by E. G. Ravenstein. London: Hakluyt Society, 1901.

Cole, F. J. *A History of Comparative Anatomy: From Aristotle to the Eighteenth Century*. London: Macmillan, 1944.

Cowper, William and Edward Tyson, "Carigueya, Seu Marsupiale Americanum Masculum. Or, The Anatomy of a Male Opossum," *Philosophical Transactions* 24 (1704–1705): 1565–75.

Cunningham, Andrew. *The Anatomist Anatomis'd*: *An Experimental Discipline in Enlightenment Europe.* Farnham: Ashgate, 2010.

Darwin, Charles. *The Origin of Species*, edited by Gillian Beer. Oxford: Oxford University Press, 1996.

Daston, Lorraine. "Curiosity in Early Modern Science." *Word and Image* 11, no. 4 (1995): 391–404.

Derrida, Jacques. *The Animal That Therefore I Am*, edited by Marie-Louise Mallet, and translated by David Wills. New York, NY: Fordham University Press, 2008.

———. *The Beast and the Sovereign: Volume I*, edited by Michel Lisse, Marie-Louise Mallet, and Ginette Michaud, and translated by Geoffrey Bennington. Chicago, IL: University of Chicago Press, 2009.

Descartes, René. *Discourse on the Method and Meditations*, translated by Elizabeth S. Haldane and G. R. T. Ross. New York, NY: Dover Publications, 2003.

Fudge, Erica. "Bad Manners at the Anatomist's Table: Tyson's Pygmie and the Naturalisation of Truth." In *The Political Subject: Essays on the Self from Art, Politics and Science*, edited by Wendy Wheeler, 14–30. London: Lawrence and Wishart Ltd, 2000.

Gesner, Conrad. *Icones Animalium*. Zurich: Christof Froshover, 1560.

Handbill. "Advertisement, June, 1968, At Moncreff's Coffee house." 2026.25 (9). British Library.

Janson, H. W. *Apes and Ape Lore in the Middle Ages and the Renaissance.* London: The Warburg Institute, 1952.

Montagu, Ashley. *Edward Tyson, M. F., M.D., F.R.S. 1650–1708, and the Rise of Human and Comparative Anatomy in England: A Study in the History of Science*. Philadelphia, PA: The American Philosophical Society, 1943.

Moxham, Noah. "Edward Tyson's *Phocaena*: A Case Study in the Institutional Context of Scientific Publishing." *Notes and Records of the Royal Society* 66, no. 3 (2012): 235–52.

Oliver, Kelly. "Elephant Eulogy: The Exorbitant Orb of an Elephant." In *The Animal Question in Deconstruction*, edited by Lyn Turner, 89–104. Edinburgh: Edinburgh University Press, 2013.

Ovid. *Ovid's Metamorphosis Englished, Mythologized and Represented in Figures by George Sandys*. Oxford: John Lichfield, 1632.

Potts, Alex. "Natural Order and the Call of the Wild: The Politics of Animal Picturing."*Oxford Art Journal* 13, no. 1 (1990): 12–33.

Raymond, Corbey and Bert Theunissen, eds. *Ape, Man, Apeman: Changing Views since 1600.* Leiden: Leiden University, 1995.

Salisbury, Joyce E. *The Beast Within: Animals in the Middle Ages*, 2nd ed. London: Routledge, 2011.

Settle, Elkanah. *Threnodium Apollinare. A Funeral Poem to the Memory of Dr. Edward Tyson.* London: Thomas Bickerton, 1708.

Smith, Justin E. H. "Language, Bipedalism and the Mind–Body Problem in Edward Tyson's *Orang-Outang* (1699)." *Intellectual History Review* 17, no. 3 (2007): 291–304.

Stow, John. *A Survey of the Cities of London and Westminster.* Vol. 1. London: A. Churchill, 1720.

Swammerdam, Jan. *Ephemeri Vita, or, The Natural History and Anatomy of the Ephemeron, a Fly that Lives but Five Hours*, translated by Edward Tyson. London: Henry Faithorne and John Kersey, 1681.

Topsell, Edward. *The Historie of Foure-Footed Beastes.* London: William Jaggard, 1607.

Tyson, Edward. *Phocæna, or The Anatomy of a Porpess Dissected at Gresham Colledge: with a Præliminary Discourse concerning Anatomy, and a Natural History of Animals.* London: Benjamin Tooke, 1680.

———. *Orang-Outang, sive Homo Sylvestris: Or, the Anatomy of a Pygmie Compared with that of a Monkey, an Ape, and a Man. To which is added, A Philological Essay Concerning the Pygmies, the Cynocephali, the Satyrs, and the Sphinges of the Ancients. Wherein it will appear that all are either Apes or Monkeys, and not Men, as formerly pretended.* London: Thomas Bennett and Daniel Brown, 1699.

———. *Orang-Outang, sive Homo Sylvestris: or the Anatomy of a Pygmie, 1699: A Facsimile with an Introduction by Ashley Montagu.* London: Dawsons of Pall Mall, 1966.

Tyson Folio. MS 618. Royal College of Physicians.

Willis, Thomas. *Two Discourses Concerning the Soul of Brutes, which is that of the Vital and Sensitive of Man.* London: Thomas Dring, 1683.

Wiseman, Susan. "Monstrous Perfectibility: Ape-Human Transformations in Hobbes, Bulwer, Tyson." In *At the Borders of the Human: Beasts, Bodies and Natural Philosophy in the Early Modern Period*, edited by Erica Fudge, Ruth Gilbert and Susan Wiseman, 215–38. Basingstoke: Macmillan Press, 1999.

Wokler, Robert. "Tyson and Buffon on the Ourang-Outang." *Studies on Voltaire and the Eighteenth Century* 155, no. 5 (1976): 2301–19.

Part III

BEYOND ORACULARCENTRISM

Chapter 6

Chris Marker's Alter Egos

The Camera and the Cat

Bonnie Gill

For all the existential musings of Chris Marker films, relatively little is known about the filmmaker himself. He was born in 1921 and died exactly ninety-one years later, on his birthday. He collaborated with other French New Wave filmmakers in the 1950s and 1960s and was a founder of a left-wing collective that produced films in the early 1970s. His filmography is extensive. But Marker valued privacy, or more accurately anonymity. When asked to appear in an official capacity Marker usually appeared in some kind of disguise; in Agnès Varda's 2008 film *The Beaches of Agnes*, Marker shows up as an orange cartoon cat named Guillaume-en-Egypte.[1] While he was still alive, Marker also maintained a cat-shaped avatar in the online world Second Life, where he conducted interviews and put up virtual installations. In at least one interview, he says he chose the name Chris Marker because it was easily pronounceable and would make it easier for him to travel,[2] but in addition to his chosen name he has been known to use pseudonyms that range from Chris Mayor to Sandor Krasna.[3] This marked preference for anonymity both contributes to and contrasts with a tendency, in his films, to capture and transmit moments in which individual identity is communicated through the exchange of glances. In several films, including *Le joli mai* (1962), *Sans soleil* (1983), and most notably *Chats perchés* (2004), Marker's camera becomes an alternate eye, a witness to the places and people he sees. The numerous cats he films, in turn, become alternate bodies that, like Derrida's cat, force a reconsideration of identity, both human and animal. These two alter egos, camera and cat, offer the ability to move in spaces to which Marker might otherwise not have access: the camera by zooming in on distant events, and the cat by masking Marker's identity as filmmaker. The camera in particular also creates points of contact with the viewer, moments in which the question of identity comes to the fore and which echo the well-known moment of eye contact that inspired Jacques Derrida's musings in "L'animal que donc je suis." Finally, despite Marker's own opacity, his films offer a distinctive way to define the self in relation to others, suggesting that by establishing points of contact with those being seen—rather than merely observing—the viewer becomes implicated in the political and social movements seen on the screen.

Marker's interest in cats, particularly their odd behaviors and steady gaze, sneaks into almost all his work.[4] As it turns out, he shares this fascination with Derrida, whose 1997 seminar "L'animal que donc je suis" explores what it means to come face to face with an animal and thereby to become ashamed.[5] Derrida finds this mutual gaze—himself, looking at the animal, while the animal looks back at his nakedness—profoundly disturbing, and hopes through his lecture to answer, in part, the question: "Whence this malaise?"[6] The cat becomes, because of its gaze, an utterly fascinating source of contemplation for the philosopher. In *Chats perchés,* Marker, like Derrida, channels his feline fascination into a philosophical inquiry into human behavior, touching on the usefulness of collective movements, the intersection of anonymity and the mutual gaze, and how art becomes an increasingly important element of political discourse in contemporary times. Generally considered a film-essay,[7] *Chats perchés* covers an assortment of protests and other popular gatherings that occurred in Paris between 2001 and 2003 while also tracing the evolution and repeated appearances and disappearances of a popular cat-shaped graffito known as M. Chat in that same city.[8] *Chats perchés* marks the epitome of Marker's unique filmmaking style, featuring a series of short narrative episodes and a proliferation of characters, none of which ever take center stage. However, unlike many of his other films, it also features a strong, though unseen, narrator, an unnamed figure whose presence is revealed through a series of intertitles that regularly appear, directing the attention of the audience by indicating possible connections between episodes, expressing more universal themes, and often making witty observations about the images on the screen. This itinerant Speaker is a noncorporeal figure made up of a highly mobile camera-eye that films what it encounters, an editorial presence that chooses and pieces together these images behind the scenes, and a disembodied voice that communicates largely via intertitle.[9] He moves through the streets without any destination in mind, keenly observing and reporting back, with commentary, on his surroundings. The Speaker also has the ability to move almost unnoticed within crowds, exposing the inner workings of group protest to the viewer's gaze. In this case, he is aided by the use of a camera that records what he sees, transmitting through image what was once put down on paper. A descendant of Walter Benjamin's roving flâneur, the Speaker remains completely anonymous, to such an extent that we never see his face or hear his voice. He is a composite figure representing a set of individuals who have a deep and powerful urge to react in some way, a desire for change, and an inability to remain quiet in the face of atrocities large and small. As a composite figure, he can be in many places and times, observing and reporting back his or her experiences and reflections. What he finds, in the film, is that hope, humor, and happiness can be found in the smallest of places: in the presence of a cat in the subway, or the smile of a woman who happens to catch his eye as she passes.[10] While Marker is not (despite his claims) a cat, nor an exact stand-in for the camera, through them he creates an elusive but identifiable alter ego (the Speaker) who searches for a source, and a balm, for the malaise felt by Parisians living under the threat of terror, war, and extremist political parties. In *Chats perchés*, both the camera and the cat become alternate ways of seeing: they roam the streets of Paris, staring unapologetically at whatever crosses their path, and allow the viewer access to the discontent expressed by the omnipresent protesters.

Many of Marker's longer and better-known films center on the political landscape of France (especially Paris) during the time of their filming. *Le joli mai* (1962), for example, revolves around the question of Algerian independence in May 1962, while *Sans soleil* (1983) questions France's place within a larger, global cultural landscape. *Chats perchés* begins by evoking the terrorist attacks of September 2001, and interrogates the various reasons that Parisians choose to congregate en masse, as well as the usual characteristics of these political and social gatherings. In particular, these tend to include some form of vocal contributions (singing or chanting), coordinated and individual motion (marching or dancing), signage (posters, flags, and banners), and the use of public space (most often streets, but occasionally green spaces). Despite the disparate causes that have gathered people together, these elements remain relatively constant across the movements included in the film. Much of *Chats perchés* is concerned with group protests, and in each case, the viewer hears and sees lengthy sequences of protesters acting as a group, often coordinating movements. Each protest features large groups of people speaking, singing, or shouting in unison, clapping, linking arms, and dancing or marching; often the group's chanting or singing is the sole soundtrack of that sequence of the film. The individuals who make up these protests are rarely identified; although the camera occasionally lingers on faces, no one person emerges to become a point of empathy for the viewer. Instead, faces are viewed in rapid succession, with the camera panning across the crowd and the crowd moving past the lens. In the opening sequence, for example, the camera often cuts off the frame at shoulder height, completely concealing the participants' identities, a framing that will later be repeated in other group shots. Other examples of camera movement that establish anonymity include the many sequences that feature only the feet of marchers in the street, or hands clapping and gesturing, and the emphasis on filming people in masks or costumes, or from the back or side. Even when faces appear, the voices of the people being filmed are often unheard, or are only heard as part of the crowd, indistinguishable on their own. Often, faces and bodies are seen through windows, physically separated from the camera not just by the glass of the lens but also that of the window. Several sequences, such as the shot of children chasing a moving light projection along the floor, are lit darkly enough that only silhouettes are visible.

Despite this tendency to film crowds, however, Marker also privileges portraiture. He often isolates individuals in the frame, centering them amidst the surroundings. There is something inherently uncomfortable about filming another; the act invokes a sense of transgression, as if viewing something uninvited, as if watching from a distance or, like Edgar Allan Poe's narrator in "The Man of the Crowd," through a smoky pane of glass. If the subject is unaware of being observed there can be no question of consent; the image has been appropriated by the filmmaker. In her essay "In Plato's Cave" Susan Sontag puts it bluntly: "To photograph people is to violate them, by seeing them as they never see themselves, by having knowledge of them they can never have; it turns people into objects that can be symbolically possessed."[11] Instead of using the camera to create or maintain distance between himself and the subjects being filmed, however, Marker establishes points of contact by allowing those he has filmed to gaze back at the camera.[12] These are the moments in which one or another of the people passing in front of the camera accidentally catches the viewer's eye, looking directly into the lens. The contact that occurs in this instant is inherently intriguing; as Marker notes in *Sans soleil*, one of the earliest rules one learns as a filmmaker is to avoid having your subjects look directly at the camera. Marker, however, relishes these moments, often replaying them in slow-motion or pausing the frame so that the eyes of the stranger remained locked in place. These moments are scattered throughout his films and are gleaned from his many travels. In each instance the fourth wall is broken and the subjects being filmed lose their distance from the viewer; although the viewer cannot literally meet their gaze, a point of contact is made, a moment of recognition is established, and the foreignness of the crowd becomes the familiarity of the intimate. The viewer becomes aware of the possibility of being seen. Once the subject gazes back, the act of viewing—of filming—becomes not an act of voyeurism but one of bearing witness—of taking some responsibility for the act of seeing by taking part, by becoming the other half. By including these points of contact, Marker suggests that the self does not exist at a distance but rather as an engaged member of the world on the other side of the lens. The films encourage recognition of the other, the stranger, as we find ourselves reflected in their gaze. They are instants—a frame, a scene, a glimpse—in which the audience sees its own reflection. As Marker himself refers to them, they are "moments that quicken the heart."[13]

Sight, and particularly the exchange of gazes, prompts many of the investigations into the contact between human and nonhuman species. For Derrida, of course, it is the cat that makes him uneasy: it is via his own reflection in his cat's eyes that he first begins to consider the questions of identity that inspired "L'animal que donc je suis." Derrida's concept of the point of contact is one of exchange, of potential equality, even co-being.[14] In true Derridian style, he puns on the double meaning of the French *suis*:

> The animal is there before me, there next to me, there in front of me—I who am (following) after it. And also, therefore, since it is before me, it is behind me. It surrounds me. And from the vantage of this being-there-before-me it can allow itself to be looked at, no doubt, but also—something that philosophy perhaps forgets, perhaps being this calculated forgetting itself—it can look at me. It has its point of view regarding me. The point of view of the absolute other, and nothing will have ever given me more food for thinking through this absolute alterity of the neighbor or of the next(-door) than these moments when I see myself seen naked under the gaze of a cat.[15]

Derrida sees himself being seen, and it startles him. We are not used to the idea of being reflected—or reflected upon—by others, and the experience can and does prompt a reconsideration of the self.

For a more literary representation, animal studies scholars often point to Rainer Maria Rilke's *Duino Elegies*, published in 1923, as including a particularly lyrical evocation of the animal gaze.[16] The eighth Elegy, which considers animal consciousness, emotion, and perspective against that of man, speaks specifically of the moment when an animal and a human face each other: "some mute animal/ raising its calm eyes and seeing through us,/ and through us."[17] Giorgio Agamben, citing from the poem, notes that

> In the eighth *Elegy* it is the animal (*die Kreatur*) that sees the open "with all its eyes," in distinct contrast to man, whose eyes have instead been "turned backward" and placed "like traps" around him. While man always has the world before him—always only stands "facing opposite" (*gegenüber*) and never enters the "pure space" of the outside—the animal instead moves in the open, in a "nowhere without the no.[18]

The idea of the *Open*—the environment and perspective specific to each species and termed *Umwelt* by Jakob von Uexküll—has since been discussed by everyone from Heidigger to Benjamin, and as Agamben makes clear man has long wondered what he might learn from contact with the animal.[19]

Jean-Christophe Bailly, in his 2007 text *The Animal Side*, echoes Derrida's conclusion: "[N]either the mountain nor the pine forest, no object and no plant, can do what any animal can do: see us, and make us understand that we are seen."[20] But Bailly also equates sight—the gaze—as a sort of *touch*; when he sees a deer emerge from the darkness as he drives down a dark road, he describes it "as if with my eyes, in that instant, for the duration of that instant, I had touched some part of the animal world. Touched, yes, touched with my eyes, despite the impossibility."[21] His chance experience leads him to conclude that "there is no supremacy, neither of humans nor of beasts, that there are only passages, fleeting sovereignties, occasions, escapes, encounters. The deer was in its night and I in mine, each of us alone. Still … I am quite sure of what I touched: … that instant opening onto another world."[22] Bailly introduces the term *point of contact* as a means to make sense of his experience both with the deer and with various other animals, referring to "the close contact, always singular and always consisting of touch, that is the ordinary mode of the bond between them and us—something scarcely formed, always nascent."[23] For him, the gaze—even if it is not returned by the animal—is a shared moment, a nonphysical but nonetheless essential connection or communication between the two beings.

Although she reflects upon a different domestic animal, the idea of communication by means of touch is also essential to Donna J. Haraway's *When Species Meet*, the guiding question of which she states as: "Whom and what do I touch when I touch my dog?"[24] Haraway refers to moments of touch as *entanglements*, *knots*, or *contact zones*, the latter in her case also referring to a specified area in dog agility trials which the competing dogs must touch with their paws before continuing on to the next obstacle.[25] She writes mainly of domestic animals, whose relationships with humans are particularly tangled, practically as well as theoretically speaking. Haraway believes Derrida did not go far enough in his consideration of the cat and that Derrida, despite his enormous powers of inquiry, was not curious enough with regard to the cat's gaze back at him. She suggests that Derrida should have considered the possibility of nonverbal communication, thus "delving into the developing knowledges of both cat–cat and cat–human behavioral semiotics when species meet."[26] Although Derrida's turn to the question of whether or not animals suffer offers valuable insights into the human conceptualization of other species, it fails—according to Haraway—to actually adopt an animalistic perspective. Derrida's relationship to his cat is thus less of a knot and more of a true point of contact.

Like Bailly, Haraway is interested in the similarities between touch and sight, particularly touch as mediated by visual technologies. Referring to a photo of a stump that looks remarkably doglike and that a friend had stumbled upon by chance, she writes: "We touch Jim's dog with fingery eyes made possible by a fine digital camera, computers, servers, and e-mail programs through which the high-density jpg was sent to me."[27] Given her terminology, she understandably considers the moment of viewing the photograph of the stump-dog as one of "haptic-optic touch."[28] Interestingly, Bailly also compares his vision of the deer in the road to "an image from a film."[29] Both Haraway and Bailly are echoing the work of Laura Marks, who in *The Skin of the Film* (2000) coins the term *haptic visuality* to refer to the "the way vision can be tactile, as though one were touching a film with one's eyes."[30] Marks uses this term as a productive entry into the possibilities for intercultural cinema, which she considers "an international phenomenon, produced wherever people of different cultural backgrounds live together in the power-inflected spaces of diaspora, (post- or neo-) colonialism, and cultural apartheid."[31] In other words, Marks is interested in the moments where one culture finds the ability to "touch" another. Where Marks and Pratt write of intersections among cultures, Bailly and Haraway address the question of those among different species.

Beyond sight and touch, one might even be able to consider contact via the sense of taste, if we stretch our tolerance of wordplay. Haraway writes:

> Among beings who recognize one another, who respond to the presence of a significant other, something delicious is at stake. Or, as Barbara Smuts put it after decades of careful scientific field studies of baboons and chimps, cetaceans, and dogs, copresence "is something we taste rather than something we use. In mutuality, we sense that inside this other body, there is 'someone home,' someone so like ourselves that we can co-create a shared reality as equals."[32]

"Delicious" is a lovely way to describe that general sense of well-being that comes from identifying a kindred spirit. But generally speaking, despite their relevance to real-world interactions among species, tasting and smelling are not (yet) useful senses when it comes to works of literature, film, or art. Nonetheless, while it is perhaps easiest to perceive our humanness in our reflection in the eyes—or the skin, taste, or smell—of animals, it is also possible to find ourselves reflected by means of nonliving entities. In particular, as Bailly notes, Benjamin asserted that even objects could in some way gaze back, writing in his study on the aura that "to perceive the aura of an object we look at means to invest it with the ability to look at us in return."[33] In fact, Benjamin is well known for associating objects and identity, having used his Arcades Project as a means of collecting and understanding Paris and its people. In walking the streets of foreign cities the visitor corresponds with his surroundings, and as countless studies on the flâneur have made clear, by collecting that which he sees the flâneur also defines himself. Charles Baudelaire describes his famous flâneur-observer in "The Painter of Modern Life" as "a mirror as vast as the crowd itself … a kaleidoscope gifted with consciousness;"[34] his M.C.G. (i.e. the painter Monsieur Constantin Guys) not only recognizes himself in his surroundings, but also in part identifies himself by so doing.

And if we can find ourselves reflected in objects, why not also in images and words? Both are mere representations of reality, but they retain a certain affective power. Harnessing the surroundings—the foreign, the alien, the not-self—in film or photography, in illustration or text, authors and filmmakers bring the other to our living rooms. This is what Marker accomplishes in *Chats perchés* by gathering images from across Paris and presenting them to our view. Marker's Speaker thus becomes the "eternal[ly] convalescent" Constantin Guys, looking for joy "amid the ebb and flow of movement, in the midst of the fugitive and the infinite,"[35] a new-world flâneur armed with a camera instead of a pen. Marker's works spring in part from these interactions with the world around him, from the people, animals, and objects he observed and by means of image and/or word transmitted to the audience. The protests filled in *Chats perchés* are full of sensuous elements, including the eye contact, the chants and other vocal elements, and hands clapping and touching each other. It is via these points of contact, these appeals to the senses, that Marker's films might best resonate with their audience.

And, lest we forget the stated premise of the film—as the intertitles put it, "où sont les chats?"—the various collectives formed by the people of Paris are paralleled by the collective of cats represented by M. Chat and his feline comrades. As with the political protests, whose effect depends on the presence of as many people as possible, as iterations of the M. Chat graffito multiply across the city they begin to take on a larger significance; they must be considered as a movement and not just as individual actions.[36] Later, the film notes, M. Chat begins to appear in other cities, including Nantes and even Zurich. The significance of this proliferation of images is itself one of the questions of the film: what do all these cat images mean, and do they correspond at all with the political and social milieu of Paris at the time? By the end of the film, the two movements have merged, as protesters begin to use images of M. Chat on signs during marches and protests. Once M. Chat appears, it becomes clear that the image has been co-opted by the movements as a symbol of liberty and peace: at least one of the accompanying signs reads: "Faites les chats, pas la guerre."

However, the cats represent more than a visual marker of liberty and peace, at least as far as the film is concerned. The camera tends to linger on the cats, even zooming in to get a closer look. They are usually visually centered in the frame, and while on the whole they share certain characteristics, many feature distinguishing details such as hats, flowers, flags, wings, even defined hands and feet. M. Chat and his clones are thus, like the dancers, an opportunity to insert a little bit of whimsy and artistic commentary into an important conversation. With their insouciant grins and bright yellow coloring, the cats demand a certain amount of irreverence. As they multiply through the city they suggest the presence of a vitality within the concrete infrastructure; they are illegal but beautiful, participating in the landscape of the city by resurfacing certain small parts of it. As they proliferate across protest signs, they signify the joy still present beneath the angry protests, a sense of peace amidst an atmosphere of war. Whether one cat with many faces, or many cats collected into one, they wander the streets, observing, smiling silently to themselves, a band of feline flâneurs. And, of course, they are always willing to make eye contact with the camera.[37]

Contemporary forms of art—and in particular, the potential intersections of artistic and political expression—thus become an alternate inquiry in *Chats perchés*. The foremost example of art within the film is, of course, M. Chat himself. The film does not delve into the details of the artistic effort, instead suggesting that M. Chat is "manifesting" himself on walls (rather than being painted by human hands.) But the graffito's status as art is highlighted by a sequence in which Marker—or rather his investigatory counterpart—attempts to establish a history of the cat figure. The tongue-in-cheek sequence argues that the smiley-faced cat appeared as far back as the Chauvet cave paintings, subsequently appearing in works by everyone from Matisse to Miró to Miyazaki. This alternative history of art is humorous, but not irrelevant; street art may not be musealized in the manner of these other works but it does belong within the lineage.[38] To support this premise this film also offers a sequence titled as a visit to the "musées de la rue" that features several wall murals across Paris as well as other street offerings, some elaborate and some as simple as the words "j'aime" or "amour" painted on the sidewalk. Further emphasizing the wide range of art forms available to a contemporary audience, the film's opening sequence, the flash mob, also reflects a newer sort of artistic movement. Flash mobs have been popularized since the early 2000s as a means for people to express themselves as a group, without necessarily having a purpose or goal in mind beyond the gathering itself, and have been considered an offshoot of performance art.[39] And Marker also films several sequences that feature street musicians performing in or around the Paris metro system, including a string ensemble, a brass band, and a folk group. Each of these sets of street musicians is featured within the frame and the film as the camera pans from instrument to instrument and their musical offerings often remaining as the soundtrack even once the camera-eye has turned to other things. Each band also gathers an audience, who are also of interest to the camera; many of them clap along or even dance.

The film thus offers examples of contemporary art in the forms of street murals, street musicians, and flash mobs. However, other art-like elements appear via the advertisements that often appear in the background of the frame and the protest chants that sustain many of the political rallies.[40] Many of the ads that appear—or, at least, earn their own place in the frame—are political ads, featuring promotions for one or another of the then-current presidential nominees. These ads, along with the countless examples that line the walls of the metro, quickly become part of the landscape of the film. In fact, the advertisements that feature players from the World Cup team become a brief focus of the film as it notes the disconnect between the size of the ads (and the salary of the players) and the failure of the team itself. The protest chants also quickly become part of the soundtrack of the film due to their almost constant presence. They tend to be rhythmic and repetitive, often featuring simple slogans and very little if any melody or tune. Even the protesters who sing the French national anthem do so with very little musicality, outpacing their accompaniment and shouting more than singing. On the whole, while the ads and protest songs are made of similar materials, feature common elements, and are used as means of expression, they are not quite art. As a result, these pairings—street art and advertisement, protest chant and street musician—call into question the very definition of art. In particular, it is interesting to look at the pairing of street art and advertisement. Both are visual, often displayed on walls and designed to appeal to the viewer's senses, but one is mass-produced and aimed primarily at selling a product, while the other is hand-painted and intended as a marker of artistic identity. Advertisements are sanctioned by the powers-that-be, while graffiti is an act of defiance. But this, of course, is part of the point: advertisements are not art but rather mass-produced art-like things. The ability to differentiate between advertisement and art, between protest song and music, between slogan and street art, and between rally and flash mob is what elevates the artist out of the crowd. Art gestates in the collective but must rise beyond it. In fact, the art that is featured in the film is not traditional or sanctioned art but rather an art of the people, art of the streets, from the brass band to the cat doodles on the walls.

Given the pairings of art and art-like projects, it is almost surprising that the film never features a clip from a commercial as a comparison to film.[41] *Chats perchés* is itself an artistic product, a fact highlighted by the many self-reflexive elements within the film. A film that is self-reflexive or that evinces reflexivity usually demonstrates an awareness of itself through a variety of techniques that include the overt intervention of the director, glimpses of the filmmaking process, asides to the viewer, or anything else that betrays the filmmaker's inherent interest in constantly questioning the mechanisms behind the fixed representation appearing on the screen. As a whole, *Chats perchés* makes little or no attempt to draw the audience into a cohesive alternate reality. There is no need to suspend disbelief in order to enter into the film, for the film itself prefers that the viewer remain aware of the film's own status as a constructed artifact. In considering reflexivity as a major impetus behind the development of ethnographic film (particularly that of Jean Rouch), Matthias De Groof highlights several of the elements that might create a particular sense of reflexivity:

> Self-interrogation and the awareness of the relativity of one's own point of view drove the ethnographic filmmaker to integrate his self-doubt into his representation of the other by means of cinematic devices. Reflexivity was translated into film language and became a cinematic device. It drove him/her also to wonder about the possibility of including the perspective of the other and his self-conception within the representation the former made of the latter, and finally to wonder about how the former would represent himself and his community through cinema.[42]

As De Groof insists, the key word to consider is "self." *Chats perchés* refers back to itself as a film regularly, often by seeming to speak directly to the viewing audience, rather than pretending it does not exist. In particular, it betrays an obvious editing of both image and sound, and a sense of humor that invites the audience to take on a certain perspective in relation to the images on the screen. Both of these are closely tied to the presence of the Speaker, whose voice—accessible through the omnipresent intertitles—gives us a glimpse of the inner workings of the film. The Speaker observes and can comment on the various clips, but also seems to have a hand in choosing them. He is therefore both inside and outside the film, both filmmaker and audience.

The opening sequence quickly establishes some of the highly noticeable camera techniques that are present throughout the film. The first evidence of the hand on the controls is the extreme zoom in on the graffito of M. Chat. The grinning cat looks down on the flash mob converging at the Centre Pompidou; it is so far distant that only an extreme zoom can even reveal its presence. The long zoom is repeated frequently, always zeroing in on a new M. Chat and in the process establishing both the surreptitiousness of the graffiti and its place within the environment. The zoom is particularly jarring given that it is often followed by closer shots of the same image; the camera eye thus reveals the conscious choice to observe M. Chat from a distance, mimicking how he (in theory) is in turn observing the French.[43] Other examples of the disconnect between sound and image: in covering the Le Pen upset, the viewer sees still shots of the people's reactions and of newspaper headlines while screams are heard in the background; a woman dancing to music at a protest is shown in slow-motion, while the music continues at full speed; M. Chat disappears, but then reappears to the sound of royal fanfare; the AIDS lie-in is paired with the opening music from a film made by one of Marker's friends and colleagues, Alain Resnais, that also deals with death on a large scale and with memory: *Hiroshima mon amour* (1959). The disconnect between sound and image asks the viewer to note these moments, and to consider what might be going on behind the scenes.[44] A particularly disconcerting sequence shows yet another shot of the men and women of Paris traveling through the subway system on a moving sidewalk, as they have been seen many times before in the film.

The soundtrack, however, has been replaced by the sound of explosions and machine-gun fire. The sequence follows the shot of Bush declaring the invasion of Iraq and Tony Blair discussing the threat of weapons of mass destruction; the explosions thus evoke the bombs that (perhaps at that very moment) were falling on Baghdad and other parts of the country. Shots of individuals and couples are intercut with black frames, and the camera pans rapidly back and forth. Finally, advertisements loom overhead, gigantic smiling faces selling something indecipherable. As with the advertisements of the soccer players who look down, Stalin-esque, on the passing crowds during the World Cup, something is clearly off. Unable to reconcile the visual and aural elements, the viewer must reflect on the process of putting the film together and on the difference between representation and reality.

Bailly suggests that points of contact with animals trigger unconscious memories, that "every animal, if we will only watch it exist and move, is the repository of a memory that surpasses the creature itself as it surpasses us, and where every instance in which its species has brushed against our own is inscribed."[45] Bailly's deer, like Derrida's cat the inspiration for a theoretical inquiry into identity, parallels many of Marker's filmed images, in which something—some unexpected moment—catches the eye and inspires a thought, a memory, a point of connection. This is the epitome of Marks' haptic moment: the eye touches the film, touches the humans, animals, objects, or images that Marker presents to our gaze. It's just a moment, just a glimpse of a deer on the road or a cat in the metro, but by taking that moment and making it durable—by including it in a film or a text—it becomes transmittable. It quickens the heart. These moments, while perhaps not as immediately accessible to our modern-day, highly attuned visual senses, nonetheless also spur recognition. In essence, by offering a kaleidoscope of reflections to our gaze they tap into the undeniable human urge to search for meaning and purpose beyond the self.

Offering these points of contact, Marker's films suggest that the viewer too is implicated in the events shown on screen. Many of the protests filmed and included in *Chats perchés* are more than mere curiosities; they are expressions of the very real dissatisfaction that exists within the French populace. In addition to the visual investigations of Paris and M. Chat, there is an intensity hidden amidst the plenitude of images, a sense that the subjects of the film—the anonymous Parisians who gather together to protest, to celebrate, to mourn, even to participate in flash mob art in the tradition of Allan Kaprow's Happenings—have a real, albeit limited, power. The subjects captured by Marker's camera are moving with a purpose, driven by an urge to do something, to react, to express emotion, to scream obscenities at a world that could permit such atrocious acts of war and terrorism as those that spur the demonstrations in which they participate. Points of contact implicate the viewer into these events, bringing the issues through the screen and into the home. Marker's alternate bodies are key to his ability to film within the crowd—the camera sees, records, and transmits, while the anonymous cat slips into the smallest of spaces and reminds us that by returning the gaze, we also reconsider ourselves.

NOTES

1. *Les Plages d'Agnès,* dir. Agnès Varda. 2008. New York: Cinema Guild, 2010. DVD.

2. Nora Alter, *Chris Marker* (Urbana: University of Illinois Press, 2006), [page #].

3. Often the credits of his films list a series of these names, some identifiably false ("Magic Marker") and some not. A partial list exists at http://www.chrismarker.ch/portraits-de-chris-marker.html.

4. *Le joli mai* (1962), *Sans soleil* (1983), *Le fond de l'air est rouge* (1977), and *Chat écoutant la musique* (1990) are good places to start; despite otherwise serious subject matter, they are full of (actual) cats. Derrida assures us, lest we try to make his cat into an analogy: "I must make it clear from the start, the cat I am talking about is a real cat, truly, believe me, a little cat. It isn't the figure of a cat. It doesn't silently enter the room as an allegory for all the cats on the earth, the felines that traverse myths and religions, literature and fables." Derrida, *The Animal That Therefore I Am*, 6.

5. Published in French as Jacques Derrida, *L'animal que donc je suis* (Paris: Editions Galilée, 2006); in English as Jacques Derrida, *The Animal That Therefore I Am*, ed. Marie-Louise Mallet, trans. David Wills (New York: Fordham University Press, 2008).

6. Derrida, *The Animal That Therefore I Am*, 4.

7. Much has been said about how to categorize Marker's films, which are neither documentary nor fiction in their structure or plot. Current scholarship has mostly settled on the term "essay film" or, alternately, "film essay." Phillip Lopate uses Marker's 1983 *Sans soleil* as an archetypal example of the essay-film in his prescriptive 1992 article, "In Search of the Centaur," noting that "Marker has an inveterate essayistic tendency that peeps out even in his more interview-oriented documentaries." Phillip Lopate, "In Search of the Centaur: The Essay-Film," *The Threepenny Review* 48 (1992): 19. Marker's work has been praised by Andre Bazin for being "unlike any other documentary" and "a truly innovative form of filmed intelligence." Nora Alter, *Chris Marker* (Urbana: University of Illinois Press, 2006), 15. Hans Richter had already invoked the idea of a film-essay in a 1940 text in terms that presage Marker's oeuvre, identifying it as a film that "produces complex thoughts that are not necessarily grounded in *reality*." Alter, *Chris Marker*, 17 (italics original).

8. The graffito began appearing in France in 1997 and eventually was used, as documented in the film itself, as a symbol of peace in the protests. Toma Vuille, *M.Chat* (Paris: Éditions Alternatives, 2010).

9. I refer to the unseen, unheard narrator-figure of *Chats perchés* as the "Speaker." There is a definite voice and personality behind this narrator but (s)he cannot be

precisely identified with Marker, though the filmmaker is almost certainly the one wielding the camera and making editorial decisions most of the time.

10. Echoes of Baudelaire's "À une passante": Charles Baudelaire *The Flowers of Evil*, trans. James McGowan (New York: Oxford University Press, 1998), 189.

11. Susan Sontag, *On Photography* (New York: Picador, 1977), 14.

12. My use of the term *points of contact* is inspired both by the idea of contact zones, especially as put forth by Mary Louise Pratt in her 1992 work *Imperial Eyes: Travel Writing and Transculturation*, and by the related terms, discussed below, established by animal studies scholars to describe the moments (often characterized by touch) when animals and humans interact. Both types of contact, between cultures and between species, involve physical as well as (or instead of) verbal communication; while the touch in this case may only be haptic, the discussion is nonetheless appropriate to a film that is, after all, about chasing cats. Pratt defines contact zones as follows: "Social spaces where disparate cultures meet, clash, and grapple with each other, often in highly asymmetrical relations of domination and subordination—such as colonialism and slavery, or their aftermaths as they are lived out across the globe today." Mary Louise Pratt, *Imperial Eyes: Travel Writing and Transculturation* (New York: Routledge, 2008), 7.

13. The quote comes from *Sans soleil* and is attributed to an eleventh century author named Sei Shonagon.

14. See also Deleuze and Guattari's section on "Becoming-Animal" from *Mille Plateaux*. Gilles Deleuze and Félix Guattari, *A Thousand Plateaus: Capitalism and Schizophrenia*, trans. Brian Massumi (Minneapolis: University of Minnesota Press, 2008).

15. Derrida, *The Animal That Therefore I Am*, 11.

16. Rainer Maria Rilke, *Duino Elegies and The Sonnets to Orpheus*, trans. A. Poulin (Houghton Mifflin Harcourt, 2005).

17. Ibid., 57.

18. Giorgio Agamben, *The Open: Man and Animal*, trans. Kevin Attell (Stanford: Stanford University Press, 2004), 57.

19. Agamben offers a lengthy discussion of the development of this term as well as other approaches to human/animal contact in *The Open: Man and Animal*.

20. Jean-Christophe Bailly, *The Animal Side*, trans. Catherine Porter (New York: Fordham University Press, 2011), 16.

21. Ibid., 2.

22. Ibid., 2–3.

23. Ibid., 5.

24. Donna J. Haraway, *When Species Meet* (Minneapolis: University of Minnesota Press, 2008), 3.

25. For a discussion of the development of the term "contact zone" via Mary Pratt, James Clifford, Juanita Sundberg, Anna Tsing, and Eduardo Kohn see Haraway, *When Species Meet*, 205–246.

26. Ibid., 56.; One approach to considering what the animal (or insect) might see was put forth by Jakob von Uexküll, who used the term *Umwelt* to postulate that

every species has its own environment, specified by its unique needs and perspectives. Agamben, *The Open: Man and Animal,* 39–47.

27. Haraway, *When Species Meet,* 5.

28. Ibid., 6.

29. Bailly, *The Animal Side*, 7.

30. Laura U. Marks, *The Skin of the Film: Intercultural Cinema, Embodiment, and the Senses* (Durham: Duke University Press, 2000), xi.

31. Ibid.

32. Haraway, *When Species Meet*, 236.

33. Bailly, *The Animal Side*, 17.

34. Charles Baudelaire, *The Painter of Modern Life and Other Essays*, trans. and ed. Jonathan Mayne (London: Phaidon Press Limited, 2003), 10.

35. Ibid., 9.

36. The originating artist was identified in 2007 as Thoma Vuille but by the end of the film M. Chat has been co-opted by others and can be seen on signs, banners, and so on. David Finnan, "Chatting with M. Chat at Paris graffiti show," *RFI.com*. 2010, http://en.rfi.fr/culture/20100316-chatting-monsieur-chat-paris-graffiti-show; Vuille, *M.Chat*.

37. "Viens, mon beau chat, sur mon coeur amoureux;/ Retiens les griffes de ta patte,/ Et laisse-moi plonger dans tes beaux yeux,/ Mêlés de métal et d'agate." Charles Baudelaire, "Le Chat," *The Flowers of Evil*, 70.

38. It also, incidentally, calls into question the validity of the image as truth, as the pictures of famous art have been rather obviously altered to include M. Chat; however, much as this topic might interest Marker, here the intent seems to be humor, not cynicism—the silliness of the premise of M. Chat resting on the lap of the Mona Lisa taking all the sting out of the potential critique.

39. See "A Flash mob Manifesto" as translated and archived at http://web.archive.org/web/20071012195306/aglomerarispontane.weblog.ro/2004-12-05/20168/Manifestul-Aglomerarilor-Spontane---A-Flashmob-Manifesto.html.

40. Unsurprisingly, several of the ads seen on screen include cats as part of their appeal to the consumer audience.

41. Actually there is one brief mention of a commercial, but it turns out to be a false one that features only a steady stream of M. Chat figures.

42. Matthias De Groof, "Rouch's Reflexive Turn: Indigenous Film as the Outcome of Reflexivity in Ethnographic Film," *Visual Anthropology* 26 (2013): 110.

43. It is fascinating to think about the larger implications of Marker also observing the crowd via his camera (which is separate from M. Chat but also a stand-in of sorts.) As of this writing I have not been able to determine who organized the flash mob in front of Centre Pompidou—could it have been Marker himself?

44. This has been widely noted in scholarship on Marker; see for example, Levine, Alison J. Murray. "Ghosts on the Skyline: Chris Marker's France after 9/11." *From Solidarity to Schisms: 9/11 and After in Fiction and Film from Outside the US,* edited by Cara C. Cilano. Amsterdam: Rodopi, 2009.; Alter, Chris Marker, 19.

45. Bailly, *The Animal Side*, 10.

WORKS CITED

Agamben, Giorgio. *The Open: Man and Animal*. Translated by Kevin Attell. Stanford, CA: Stanford University Press, 2004.

Alter, Nora. *Chris Marker*. Urbana, IL: University of Illinois Press, 2006.

Bailly, Jean-Christophe. *Le Versant animal*. Paris: Bayard, 2007.

———. *The Animal Side*. Translated by Catherine Porter. New York, NY: Fordham University Press, 2011.

Baudelaire, Charles. "Le Peintre de la Vie Moderne." In *Oeuvres Complètes de Charles Baudelaire,* edited by Jacques Crépet Vol. 3, 49–110. Paris: Louis Conard, 1925.

———. *The Painter of Modern Life and Other Essays*. Translated and edited by Jonathan Mayne. London: Phaidon Press Limited, 2003.

Benjamin, Walter. *Charles Baudelaire: A Lyric Poet in the Era of High Capitalism*. London: Verso, 1983.

———. *The Arcades Project*. Translated by Howard Eiland and Kevin McLaughlin. Cambridge, MA: The Belknap Press of Harvard University Press, 1999.

Chats perchés (The Case of the Grinning Cat). Directed by Chris Marker. 2004. Brooklyn, NY: Icarus Films, 2008. DVD.

Cooper, Sarah. *Chris Marker*. Manchester: Manchester University Press, 2008.

Corrigan, Timothy. *The Essay Film: From Montaigne, After Marker*. New York, NY: Oxford University Press, 2011.

Cowie, Elizabeth. "Specters of the Real: Documentary Time and Art." *Differences* 18, no. 1 (2007): 87–127.

De Groof, Matthias. "Rouch's Reflexive Turn: Indigenous Film as the Outcome of Reflexivity in Ethnographic Film." *Visual Anthropology* 26 (2013): 110.

Deleuze, Gilles. *Cinema 1: The movement-image*. Translated by Hugh Tomlinson and Barbara Habberjam. Minneapolis, MN: University of Minnesota Press, 2009.

———. *Cinema 2: The time-image*. Translated by Hugh Tomlinson and Robert Galeta. Minneapolis, MN: University of Minnesota Press, 1989.

Deleuze, Gilles and Felix Guattari. *A Thousand Plateaus: Capitalism and Schizophrenia*. Translated by Brian Massumi. Minneapolis, MN: University of Minnesota Press, 2008.

Derrida, Jacques. *L'animal que donc je suis*. Paris: Editions Galilée, 2006.

———. *The Animal That Therefore I Am*, Edited by Marie-Louise Mallet. Translated by David Wills. New York, NY: Fordham University Press, 2008.

Haraway, Donna J. *When Species Meet*. Minneapolis, MN: University of Minnesota Press, 2008.

Hiroshima mon amour. Directed by Alain Resnais. 1959. New York: The Criterion Collection, 2007. DVD.

Le joli mai (The Lovely Month of May). Directed by Chris Marker. 1962. Issy-les-Moulineaux: ARTE France, 2013. DVD.

Lopate, Phillip. "In Search of the Centaur: The Essay-Film." *The Threepenny Review* 48 (1992): 19–22.

Lupton, Catherine. *Chris Marker: Memories of the Future*. London: Reaktion Books, 2005.
Marker, Chris. *A Grin without a Cat*. London: Whitechapel Gallery, 2014.
———. *La Jetée: ciné-roman*. Brooklyn, NY: Zone Books, 1992.
———. *Staring Back. Columbus: Wexner Center for the Arts*, Columbus, OH: The Ohio State University, 2007.
Marks, Laura U. *The Skin of the Film: Intercultural Cinema, Embodiment, and the Senses*. Durham, NC: Duke University Press, 2000.
Naficy, Hamid. *An Accented Cinema: Exilic and Diasporic Filmmaking*. Princeton, NJ: Princeton University Press, 2001.
Neupert, Richard. *A History of the French New Wave Cinema*. Madison, WI: University of Wisconsin Press, 2002.
Nichols, Bill. *Representing Reality: Issues and Concepts in Documentary*. Bloomington, IN: Indiana University Press, 1994.
Poe, Edgar Allan. "The Man of the Crowd." In *The Complete Tales and Poems of Edgar Allan Poe*, 414–420. New York, NY: Race Point Publishing, 2014.
Pratt, Mary Louise. *Imperial Eyes: Travel Writing and Transculturation*. New York, NY: Routledge, 2008.
Rascaroli, Laura. "The Essay Film: Problems, Definitions, Textual Commitments."*Framework: The Journal of Cinema and Media* 49, no. 2 (2008): 24–47.
Rilke, Rainer Maria. *Duino Elegies and the Sonnets to Orpheus*. Translated by A. Poulin. Boston, MA: Houghton Mifflin Harcourt, 2005.
Rivoire, Annick. "Interview: Chris Marker et M. Chat: Chats discutent."*Liberation*. 2004. http://next.liberation.fr/arts/2004/12/04/chats-discutent_501770.
Sans soleil (Sunless). Directed by Chris Marker. 1983. New York: The Criterion Collection, 2007. DVD.
Sontag, Susan. *On Photography*. New York, NY: Picador, 1977.
Vuille, Thoma. *M.Chat*. Paris: Éditions Alternatives, 2010.
Warner, Rick. "Go-for-broke games of history: Chris Marker between 'old' and 'new' media."*Post Script* 29, no. 1 (2009): 14–27.

Chapter 7

Scenting Wild

Olfactory Panic and Jack London's Ocular Dogs

David Huebert

Nonhuman-oriented thinking has long been preoccupied by the visual sensorium. Many leading animal studies texts rely on tropes and logics of vision—John Berger asks "Why Look at Animals?,"[1]Anat Pick interrogates Temple Grandin's idea of "thinking in pictures,"[2] Jacques Derrida reflects on being "caught naked, in silence, by the gaze of the animal."[3] Of course, such works use the sensory field of vision to probe and blur the membranes of species being in productive ways, and critics who emphasize the visual are most often aware of the pitfalls of occluding other senses. But the reliance on sight as governing metaphor remains a troublesome methodology with which to engage nonhuman animals, most of whom do not rely on sight to the same extent as humans. Strangely, the thinkers listed above, as well as many others, rarely engage with the olfactory sensorium,[4] which we know to be a, if not the, central form of nonhuman being and knowing. This essay encourages a shift in creative and critical approaches to literary animals—a turn away from the visual, down the sensually queer path of the olfactory, and towards the diverse spectrum of intersecting and overlapping sensory machineries that inhabit the far side of the homogenizing paradigm of the visual. Beginning by examining Jack London's striking neglect of smell in his 1903 literary foray into canine experience, *The Call of the Wild*, this essay proceeds to unpack a larger theoretical paradigm—from early sexology to modern animal studies—that has consciously and unconsciously marginalized smell as a pejoratively bestial vehicle of perception. In the cultural and intellectual history I examine here, smell has been simultaneously eroticized and ghettoized as an animalistic sense. I conclude by examining some recent scholarship and literature that seeks to queer current sensory norms by opening new conduits for thinking and writing the olfactory.

LONDON'S OCULAR DOGS

The Call of the Wild has become one of the canonical texts of literary animal studies. However, in the large body of research on dogs in London's fiction, astonishingly little attention has been paid to the nonvisual sensorium in his work.[5] London's highly ocular writing leads naturally to a critical emphasis on vision. Despite London's claim to authentically render canine experience, narratives such as *White Fang* and *The Call of the Wild* offer highly inadequate portrayals of the canine olfactory. While London, writing through Buck's eyes, offers a luxuriant feast for the visual, he mentions smell only in passing, thus leaving a precarious lack at the heart of these stories. London's dog tales and the criticism surrounding them thereby depend on and reinscribe the damaging tendencies of what Martin Jay, in his groundbreaking 1993 work *Downcast Eyes*, calls the "ocularcentric."[6] My central argument in this section is that London's resistance to smell demonstrates a limitation of his zoopoetics and is symptomatic of his immersion in a larger ocularcentric paradigm that relegates smell to the murkiest closets of the human sensory echelon. London's reluctance to portray canine smell has its roots in the sexological association of smell with queerness and perversion, a connotation still present in cultural discourses, including animal studies, today. London is by no means unique as an artist caught up in the ocularcentrism of generic paradigms such as realism and naturalism; but his claim to offer an authentic account of the canine sensory experience makes him a particularly relevant (and vulnerable) object of analysis.

London was eager for his dog tales to offer realistic and scientifically accurate portrayals of canine consciousness. In a February 1903 letter to his publisher George Brett, London anxiously defends his work-in-progress, *The Call of the Wild*: "It is an animal story, utterly different in subject and treatment from the rest of the animal stories which have been so successful."[7] London, here, is suspiciously eager to differentiate his own work from the emergent subgenre Lillian Carswell calls "sentimental animal stories,"[8] narratives that had "consumed the reading public" in the years leading up to the publication of *The Call of the Wild*.[9] American readers, to whom London was catering, were much taken by animal novels such as Anna Sewell's *Black Beauty*, published in an American edition in 1890. Other popular animal writers included Ernest Thompson Seton and Charles G.D. Roberts, both of whom were, like London, involved in the nature fakers controversy that began in 1903.[10] Tales like *The Call of the Wild* and *White Fang* offered a marketable masculine rugged twist on what had by 1900 become an established and beloved canon of relatively saccharine literature about pets and wild animals. London explicitly elaborates his formal prerogatives in his 1906 essay, "The Other Animals," in which he affirms the scientifically informed realism of his stories: "I endeavored to make my stories in line with the facts of evolution."[11] Here he distinguishes his own work from the "'humanizing' of animals" of which "several of the 'animal writers' have been profoundly guilty,"[12] again differentiating himself from the sentimental literary pack in the name of a scientific verisimilitude he believed would help to combat anthropomorphism. But what does it mean to aspire to realism, an inherently anthropographic and enormously visual genre, when engaging nonhuman consciousness? What London, perhaps unwittingly, aspires to in *The Call of the Wild* and *White Fang* is a visual realism that occludes the crucial centrality of smell to canine being and knowing.

Biological and ethological research affirms the importance of smell for many nonhuman animals, including dogs.[13] Behaviorist Temple Grandin states that dogs "smell things at a threshold a thousand times lower than people can,"[14] suggesting that olfaction shapes canine perception to an extent humans would naturally have difficulty conceiving. Dog trainer Vicki Hearne uses an evocative metaphor to compare human and canine perception: "[f]or dogs, scenting is believing. Dogs' noses are to ours as a map of the surface of our brains is to a map of the surface of an egg."[15] As a committed follower of Darwin,[16] London would have been aware of the prowess of the canine olfactory sensorium. In *The Descent of Man, and Selection in Relation to Sex* (originally published 1874), Darwin writes that "the sense of smell is of extremely slight service" to human beings,[17] as opposed to "animals which have this sense highly developed, such as dogs and horses."[18] Darwin knew well that the canine sense of smell was vastly superior to the clumsy human olfactory system, and London's familiarity with the science of his day would suggest he also knew as much. Presumably, attempting to portray such a vastly different sensory makeup would have been an exciting challenge to a writer who was, like London, seeking to fictionally manhandle the canine corpus. But the depiction of smell remains one of the major weaknesses of London's attempt to dramatize "scientifically plausible and realistic representations of mental phenomena in his animal protagonists."[19] As I will demonstrate shortly, London persistently occludes the olfactory, searching, instead, for the homosensory utopia of a panoptic visual realism.

One limitation to this line of analysis is that there is a problem in the very idea of isolating a single sense in analyses of representations of organic bodies. As Classen, Howes, and Synnott note, "olfactory symbol systems" always "exist and acquire meaning within a multisensory context."[20] Like vision and hearing, smell does not exist in isolation but as part of the confused, multivalent technics of embodied being. However, the complex interactions of multisensory embodied being itself offers an argument for paying more attention to smell. While the current ocularcentric paradigm has done its best to sanitize sensory being into the monosensorium of the visual, an increased focus on smell—particularly as it works in conjunction with touch, sound, sight, and taste—enables a more accurate awareness of the vast complexities of human and nonhuman multisensory experience. It is not my intention to invert the established sensorium, prioritizing smell over vision. But I would like to suggest that a more honest attendance to human and nonhuman sensory experience would require greater acknowledgement of the robust role of smell in lived bodily experience.

London's omission of smell emerges most prominently in a trifecta of chapters in *White Fang* beginning with "The Gray Cub." Here, London attempts the enormously ambitious exercise of offering a detailed phenomenological delineation of a wolf-dog coming to know its sensory world. Born into a world of murk, "the gray cub" soon becomes fixated on the "wall of light" at the mouth of his cave.[21] London describes the cave's entrance thus: "It had been an irresistible attraction before ever his *eyes* opened and *looked* upon it. The *light* from it had beat upon his sealed *lids*, and the *eyes* and the *optic nerves* had *pulsated* to little, spark like *flashes*, warm-*colored* and strangely pleasing."[22] The cave mouth, according to this description, brings nothing inside except what is available to visual apprehension. Buck's consciousness in *The Call of the Wild* is similarly driven by the ocular: "Buck *saw* a slipshod and slovenly affair … also, he *saw* a woman. Mercedes the men called her. … Buck *watched* them apprehensively as they proceeded."[23] This entire scene, ostensibly oriented through Buck's sensorium, takes place on the level of the visual. There is not even an aside for how this woman might smell differently from the strictly male bodies Buck has been encountering throughout his sojourn through the wilderness. London claims to overcome "homocentrism" and describe the canine bodily experience according to the science of evolution, but his flagrant prioritization of the ocular compromises this project. London, in fact, perpetuates one of the most insidious and ubiquitous symptoms of anthropocentrism; he caters to a familiar sensory hierarchy, cascading in importance from the most human (sight) to the most bestial (smell).[24]

Attempting to represent canine consciousness in literature is an ambitious endeavor, and London's project is no doubt driven by genuine and earnest curiosity. I do not mean merely to wag the proverbial tail at London for trying to fictionalize canine experience. Literature can never accomplish everything a given reader wants it to achieve, and the sympathetic reader can forgive London his ocularcentrism because of the simple fact that the human lexicon of smell is enormously lackluster.[25] London certainly was not alone in his aversion to smell, and the next section of this essay examines the larger cultural and theoretical paradigm that would have encouraged him to occlude smell—a raunchy, animalized, and queerly pathologized sense—from his literary excursion into canine *umwelt*.

OLFACTORY PANIC

London's occlusion of smell from his dog stories may well have derived from an aversion to the perceived, threatening animality of olfaction, a sense historically associated with the genitals and anus, the pungent and unsightly underbits of civilized literary culture. The anus is, of course, a locus of profound canine interest and joy. Harlan Weaver notes that "most dogs live in what could be termed a 'butt culture'": "Butts and smells are not only a big part of dog life but also central to how dogs come to know the worlds and cultures, canine and human, in which they participate."[26] London's reluctance to enter the world of smell, then, may be at least in part a phobic withdrawal from the anal engagements required by an authentic exposure to canine experience. The author approaches the canine psyche as far as he can go, but, panicked, retreats at the threshold of the posterior and the sense with which it is most closely associated.

London's early-twentieth-century world understood smell as an unpleasant and uncivilized vestige of humanity's bestial past. This was a culture, as Dana Seitler notes, steeped in a "new scientific visual order" spurred by Louis Daguerre's invention of photography in 1839.[27] The emergent visual paradigm, developing alongside the discovery of the fossil record and increasingly plausible hypotheses about deep time and humanity's atavistic past, also sought to diminish the importance of smell and odor for the human animal. Though biological research now shows that smell plays a crucial function in human social interactions such as recognition and mate choice,[28] in the early twentieth century the visual-scientific paradigm sought to relegate smell to the darkest closets of the human species' primitive past. Irene Tucker writes that "in the final quarter of the eighteenth century, *skin* suddenly came to be privileged as the primary sign of racial identity."[29] Photography and other contributions to the ocular scientific paradigm of this period helped to concretize the drive to mark nonwhite otherness as legible to the visual register.

Freud is perhaps the most important figure in the Western intellectual history of smell, and the work of sexology and psychoanalysis played crucial roles in shaping the public perception of smell as an inferior, animalistic sense. Along with Charles Darwin and early sexologists Havelock Ellis and Richard Von Krafft-Ebing, Freud championed the "powerful denigration of smell by Europe's intellectual elite."[30] For Freud, the raising of the human head at some point deep in the evolutionary past signals a "devaluation of the olfactory stimuli" that makes pleasures such as the "pungent genital odours" vulgar and obsolete to the civilized,[31] ocular human subject. Case studies such as "The Rat Man" (1909) and "The Wolf Man" (1918) demonstrate that Freud feared the threat of animality and that he associated that threat with smell, which he also linked to "perversions" such as anal eroticism. In his study of the Wolf Man, Freud writes that "a sexual preference ... for the hind parts of the body is a general characteristic of people who are inclined to an obsessional neurosis ... It is part of the fabric of the anal-erotic disposition and is one of the archaic traits which distinguish that constitution."[32] Freud, here, not only deems that sexual interest in buttocks is pathological, he associates that pathology with "archaic" or primitive dispositions—erotic interest in hindquarters, in other words, signifies arrested development at the level of the species.[33] In a speculative tangent at the end of "The Rat Man," Freud more explicitly draws the link between smell, animality, and neurological pathology. Here he notes that his patient, the Rat Man, was subject to "osphresiolagnia,"[34] a pathology characterized by sexual interest in smells. Speculating on this condition, Freud asks if the "repression of his osphresiolagnia may not have had a considerable share in [humanity's] susceptibility to nervous disease."[35] Remarkably, Freud here wonders whether the human species' ascension up onto the hind legs and the subsequent turn away from olfactory eroticism might in fact have caused the proliferation of human neurological disorders he encountered in the modern populace. In other words, a more mentally healthy creature might be one that engaged more deeply with smell. But Freud does not advocate a return to smell. Though he tends to remain neutral regarding the question of evolutionary progress and does not necessarily use "primitive" in a pejorative sense, Freud maintains that olfactory eroticism, like anal eroticism, is a vestige of the primordial human past.

Freud surely derived some of his thoughts on smell from early sexologists such as Havelock Ellis and Richard Von Krafft-Ebing, both of whom perceived olfaction as a marker of primitive sexuality and see erotic interest in genital odor as perverse. Writing in 1906, Ellis baldly states that smell "is a primitive sense which had its flowering time before men arose."[36] Similarly, in his *Psychopathia Sexualis* (1886) Krafft-Ebing writes that although for "beasts the influence of olfactory perception on the sexual sense is unmistakable," he is "inclined to doubt" whether the same is true in humans "under normal conditions."[37] For both writers, sexually normal humans, unlike animals, are not aroused by olfactory stimulation. Ellis goes so far as to explicitly associate olfactory eroticism with male and female "inverts,"[38] giving smell an important sensory location in the sexological continuum of abnormality. For Freud, smell and anal eroticism are conjoined markers of the primitive, but, as Greg Forter notes, the Freudian model also relies on an imaginative prehistoric sensorium that does not "*distinguish between the anus and the genitals*."[39] In psychoanalysis, the penis/scrotum and, more sensationally, the vagina—which Freud deems to be "uncanny" in his essay of the same name[40]—take up a semantic position adjacent to the anus as markers of olfactory phobia. Certainly, for Ellis, olfactory perversion also finds a scapegoat in the vagina;[41] citing a patient who "dislike[s] the smell of a woman's vagina," Ellis quips that this "statement seems to express the feelings of many if not most men."[42] Insofar as the animality of smell produces an olfactory unease among early sexologists, that anxiety has grounds both in anal phobia and a version of what Susan Fraiman has called "pussy panic."[43] The norm, sexologists wanted their readers to believe, was to find human anal and genital body odor repellent rather than attractive. Such examples demonstrate that the history of sexology and its cousin psychoanalysis sought to establish smell as an animalistic, abnormal, and threateningly erotic force. Smell, here, not only signifies the animal. Smell is the animal. Smell is the "beast within," the animal of pathologies and sexual perversions, the animal of "inversion" and anal eroticism, the boogeyman of early sexology. Placed within this cultural context, smell begins to waft unpleasantly out from the sensory closet of Jack London's writings, a conspicuous odor soiling his visually sanitized search for a scientifically accurate depiction of canine consciousness.

The turn away from smell came about alongside the large-scale urbanization of the late nineteenth century—the creation of the city as an entirely "human" realm. It is by no means incidental that the historical denigration of smell as a bestial sense to be cordoned off from civilization took place alongside widespread urbanization and the concurrent removal of meat-processing technologies from the areas where citizens were to be fed by animal bodies that were becoming increasingly unrecognizable as such. Nicole Shukin offers an instructive account of the role of smell in the development of urban-industrial meat-production technologies:

> As slaughter and rendering were turned into mass operations in the nineteenth and twentieth centuries … [m]odern renderers became acutely conscious of olfactory leakage from the industrial cooking of animal remains and of a populace whose senses risked being offended by reminders of a grisly business exiled to the margins of public consciousness.[44]

Shukin deploys noticeably suggestive language to make this important point regarding the removal of meat production from the public eye. What was being hidden at this time was not only the "unsightly" smells of living and dying animal bodies. There was also, as the work of Freud and his colleagues demonstrates, a concurrent desire to repress the animality of the human body, an animality thought to be connected to smell. Suggestive phrases such as "olfactory leakage" demonstrate the extent to which the animal body is in part threatening as a purveyor of pungent odors that are fearfully erotic. The push towards urbanization and the sanitation of smell takes place concurrently at the level of the city and the site of the human sexual organism. In both cases, smell must be cast out of sight, banished to the peripheral shadows of the bestial. Writing about smell, therefore, might have been seen as something of a perverse act on Jack London's behalf, a troubling sign of his own interest in a primitive, sexualized sense. This would have been a crass, dirty subject to approach. Had he discussed at length the smell of animal anuses and dogs in heat, London's books no doubt would not have sold as well or attained their canonical status. London's fiction remains polite and mannered, a vestige of its *fin-de-siècle* milieu. Writing the animal accurately, if such a thing can be done, surely requires the writer to get a little more dirty than this. London, limited by the sexual hang-ups of his era, remained unable to tackle the anal-animality of the olfactory.

To what extent is animal studies today still burdened by psychoanalytic olfactory phobia? Animal studies, like psychoanalysis,[45] takes a visual spectacle as its "primal scene."[46] Jacques Derrida's well-known encounter with his cat in *The Animal That Therefore I Am*. Here Derrida not only blatantly occludes smell in his exegesis of a visual interspecies exposure, he also offers a howling allusion to the work of Jacques Lacan when he meditates on "the mirror of a shame ashamed of itself."[47] Drawing on Lacan's well-known essay "The Mirror Stage as Formative of the *I* Function as Revealed in Psychoanalytic Experience," Derrida here enacts a mirror stage at the level of interspecies being. This is a psyche-foundational moment with which most readers can identify: looking at a nonhuman animal and recognizing the animal looking back, taking stock, exposing the precariousness and vulnerability of a fragile species identity. In an inversion of Lacan's mirror stage—in which the child for the first time encounters its separation from the external world—what Derrida's encounter enacts is the experience of realizing one's existence beyond human exceptionalism, as part of the vast continuum of nonhuman life. This moment of profound exposure—where the human stands precarious, naked, and vulnerable before the nonhuman gaze—has been enormously generative for many readers of Derrida's work. But it is also important to note that the visual enframing of the moment situates it in the profoundly human terms of ocularity. Though Derrida himself has been a long-time critic of the phallogocentrism at work in Lacan's "scopic field,"[48] here he, like London, attempts to engage a nonhuman animal without considering smell. Cats, like dogs, communicate in an enormously complex olfactory language.[49] Unlike humans and dogs, cats also have very limited use of facial expressions.[50] So why is Derrida so concerned with looking into the face of this cat, and why does he eschew smell altogether? While Derrida primarily notices the gaze of the cat, the cat might mark the strange chemical smell of shampoo or the bizarre compulsive human desire to remove body odor by showering. Donna Haraway warns against the "basically imperialist" tendency underlying the literary "fantasy of climbing into heads,"[51] a reprimand which would seem to censure both London and Derrida. Perhaps, though, the inevitable shortcoming of literary and philosophical attempts to empathetically inhabit animal perception involves a human overvaluation of brains and intellect. Maybe "climbing into heads" is the wrong fantasy because humans could learn more by climbing into noses and attempting to access the stinkier areas of animal experience.

Diagnosing the "pussy panic" at work across animal studies,[52] Fraiman takes Cary Wolfe and his male colleagues to task for coronating Derrida as the "founding father" of animal studies while ignoring the work of female field-pioneers such as Carol Adams,[53] Vicki Hearne, and Jane Goodall. Fraiman's thesis is that "men working in the area of contemporary animal studies—men siding with animals—may indeed feel threatened by "castration." Proximity to this feminized realm may even induce a degree of species/gender anxiety I am tempted to call, with a nod to Eve Sedgwick, pussy panic."[54] I would like to suggest that a correlate of Fraiman's "pussy panic" is a troublesome focus on the ocular throughout animal studies, a focus that occludes the beastly possibilities of smell. If researchers such as Wolfe are wary of the gendered emasculations involved in taking up the cause of animals, animal studies practitioners more generally may anxiously avoid serious discussions of smell precisely, if unconsciously, because of its historical associations with a so-called perverse animality.

While he may be subject to a certain "pussy panic," Derrida does not shy away from olfaction and the animality associated with it. First, he rightly notes that "this zone of sensibility is [. . .] neglected or reduced to a secondary position in philosophy and in the arts."[55] Furthermore, in a key moment in *The Animal That Therefore I Am*, he suggests that smell might serve as a model for new, creative ways of thinking: "The steps to be consistently followed [*cette démarche suivie*] will indeed have to resemble those of an animal seeking to find or seeking to escape. Do they not resemble the running of an animal that, finding its way on the basis of a scent or a noise, goes back [. . .] to sniff the trace of another."[56] What Derrida advocates, here—and he is speaking both about his project in this late lecture and about the larger task of philosophy—is a turn towards a thinking that imitates smell. This cultivation of olfactory thinking gains momentum throughout *The Animal That Therefore I Am* as the classic Derridean concept of the trace merges with an olfactory notion of tracking or following; Derrida insists that "one should be able to sniff the trace" of language and adds that it is essential "to track, to sniff, to trail, and to follow" the movements of thought.[57] The trace, here, becomes an olfactory track, a path to be pursued in animalistic fashion. At least since *Of Grammatology* (first published in French in 1967), Derrida advances a concept of the trace as a crucially ephemeral presence: "The trace *is nothing*, it is not an entity, it exceeds the question *What is?* and contingency makes it possible."[58] Of the senses available to humans, what could embody this contingent, palpable nothingness more than smell, a sense that is pungent and visceral yet unarchivable, always fleeting. Smell can be enormously vivid—it can evoke highly emotional memories, can cause us to vomit or faint. It is the most visceral, immediate, and immanent of the senses. But a smell is never a vista; it cannot be recorded; it is only ever a whiff, a trace. Smell, as a late iteration of foundational Derridean thinking, also allows a reappraisal of the tired regime of ocularcentrism and provides a sensually queer alternative to the traditional normative hierarchy of the senses.

QUEERING THE OCULAR

Since the time of London's writing, many literary works have attempted to evoke the canine perspective. Notable examples include T.C. Boyle's short story "Dogology" (2002), Dave Eggers' tale "After I Was Thrown in the River and before I Drowned" (2004), and Bruce Cameron's novels *A Dog's Purpose* (2010) and *A Dog's Journey* (2012). Among recent fiction written from the canine point of view, André Alexis' remarkable novel *Fifteen Dogs* (2015) stands as a singular achievement in part because of its bold eschewal of the realist agenda. Unlike London, whose project was limited by a commitment to visual-scientific realism, Alexis shrugs off the incitement to verisimilitude and thereby opens productive room in which to cultivate interspecies formal play.

Alexis begins his novel in mock-epic fashion with a council of the gods. Hermes and Apollo meet in a downtown Toronto pub and debate whether human intelligence is a "gift" or an "occasionally useful plague" and decide to settle the matter by endowing a pack of dogs with human intellect.[59] Establishing this conceit at the outset allows the reader to suspend skepticism as the author forfeits the claim of scientific accuracy, creating a mythopoetic story-world in the manner of Greek drama. Rather than, like London, seeking to overcome the human perspective guiding his fiction, Alexis acknowledges what Marco Caracciolo, in a reading of Kafka's fictional explorations into canine consciousness, calls "the ineliminable anthropocentrism of human representations of animal experience."[60] Instead of trying to achieve a plausible and scientifically accurate portrayal of canine consciousness in a language and literary form that is always inescapably human, Alexis skirts accuracy and authenticity, opting, instead, for something closer to what Canadian nature poet Don McKay calls "[a]nthropomorphic play."[61]

As Alexis' novel demonstrates, literature and criticism have recently begun to excavate some of the previously buried literary potentials of smell. Since the 1993 publication of Jay's *Downcast Eyes*, there has been an emergent body of scholarship seeking to creatively reconsider the previously ghettoized role of the olfactory in Western culture.[62] Much of this work insists that attention to smell may have the potential to trouble or at least confuse hierarchies of race and species being that were calcified in the visual-scientific paradigms of the late enlightenment. A bold critic on the role of smell in diasporic literature, Stephanie Oliver writes that because "smells' very existence depends on mixing and blending," olfaction "has the potential to subvert Eurocentric notions of pure origins and rigid boundaries."[63] While smell is a confusing sense for human perception, Oliver suggests its conceptual power might lie precisely in that confusion. Environmental philosopher Michel Serres writes that French is a "wise and true language" partly because it "calls the exhalation of a fragrance a *bouquet*."[64] The productive potential of smell in Oliver and Serres' accounts lies in its capacity to produce mixture and confusion rather than coherence. Similarly, Mark Graham notes that while the "dominance of the visual in Freud and Lacan's work is inseparable" from their constructions of the "heteronormative order,"[65] smell entails crucially "transient and elusive qualities" that align the olfactory with poststructuralist and queer-theoretical destabilizations of meaning.[66] Smell, for Graham, becomes a notably queer sense.[67] Certainly, olfaction provides a ready and under-deployed companion for theorists seeking to debunk the conjoined Enlightenment regimes of reason, visuality, racial purity, and normative gender and sexuality. But there is also something queerly subversive about smell beyond the level of human sexuality: smell is both the most erotic and the most animalistic of senses. How might humans and nonhuman animals pass in and out of one another differently if we cultivate the erotic *anima* of smell? If smell and animality are, as I have argued here, part of a yoked matrix of animal-sexual repression, what possibilities for intraspecies and interspecies *eros* may emerge by unleashing the olfactory beast within?

Of course, no one sense exists in isolation. What the rehabilitation of smell may offer, then, is not just new attention to a traditionally marginalized sense but a new way of thinking of the bodily sensorium more generally as a nexus of overlapping points of contact where sensory experience pulsates fluidly—embodied experience not as a hierarchy of competing senses but as a wavering constellation of sensory cross-pollinations. Smell has the potential to become a portal, an exposure where human sensory experience can be reanimated, reviscerализed, and differently eroticized. What better place to begin to queer and disrupt the normative structures of the traditional human sensorium?

NOTES

1. John Berger, "Why Look at Animals?" In *The Animals Reader: The Essential Classic and Contemporary Writings*, ed. Linda Kalof and Amy Fitzgerald (New York: Berg, 2007).

2. Anat Pick, *Creaturely Poetics: Animality and Vulnerability in Literature and Film* (New York: Columbia University Press, 2011), 66.

3. Jacques Derrida, *The Animal That Therefore I Am*, trans. David Wills (New York: Fordham University Press, 2008), 3.

4. One animal studies practitioner who does address smell is Nicole Shukin, whose powerful book *Animal Capital: Rendering Life in Biopolitical Times* (Minneapolis: University of Minnesota Press, 2009) includes sustained discussions of the role of smell in industrial meat rendering (63–66; 93–96). But Shukin is the clear exception to the rule. Here is a cursory list of major animal studies texts that do not include an entry for "smell" in their indices or sustained discussions of the olfactory: *Animal Stories: Narrating Across Species Lines* by Susan McHugh (Minneapolis: University of Minnesota Press, 2011); *Animal Rites: American Culture, The Discourse of Species, and Posthumanist Theory* by Cary Wolfe (Minneapolis: University of Minnesota Press, 2003); *French Thinking About Animals* edited by Louisa Mackenzie and Stephanie Posthumus (East Lansing: Michigan State University Press, 2015); *The Animal Studies Reader* edited by Linda Kalof and Amy Fitzgerald (New York: Berg, 2007); *When Species Meet* by Donna J. Haraway (Minneapolis: University of Minnesota Press, 2008); *Creaturely Poetics: Animality and Vulnerability in Literature and Film* by Anat Pick (New York: Columbia University Press, 2011); *Melancholia's Dog: Reflections on Our Animal Kinship* by Alice Kuzniar (Chicago: The University of Chicago Press, 2006); *Loving Animals: Toward a New Animal Advocacy* by Kathy Rudy (Minneapolis: The University of Minnesota Press, 2008); *Animalia Americana: Animal Representations and Biopolitical Subjectivity* (New York: Columbia University Press, 2013); *The Birth of a Jungle: Animality in Progressive-Era U.S. Literature and Culture* by Michael Lundblad (New York: Oxford University Press, 2013).

5. For example, Jonathan Auerbach, Lillian Carswell, and S.K. Robisch all ignore smell in their analyses of London.

6. Martin Jay, *Downcast Eyes: The Denigration of Vision in Twentieth-Century French Thought* (Berkeley: University of California Press, 1993), 3. Jay is, of course, the central figure in recent investigations into the olfactory. My work here owes a huge debt to his thinking, and anyone interested in this subject should certainly consult *Downcast Eyes*.

7. Earle Labor, Robert C. Leitz, III, and I. Milo Shepard, *The Letters of Jack London: Volume 1: 1865–1905* (Stanford: Stanford University Press, 1988), 342.

8. Lillian Carswell, "'The Power of Choice': Darwinian Concepts of Animal Mind in Jack London's Dog Stories," in *America's Darwin: Darwinian Theory and U.S. Literary Culture*, ed. Tina Gianquitto and Lydia Fisher (Athens, GA: The University of Georgia Press, 2014), 306.

9. Ibid., 306.

10. The controversy, between naturalists such as John Burroughs and USA President Theodore Roosevelt on the one side and writers such as London, Ernest Thompson Seton, and William J. Long on the other, receives extensive treatment by Ralph H. Lutts in his history, *The Nature Fakers: Wildlife, Science, and Sentiment* (Golden, CO: Fulcrum, 1990). The fiery public debate began in March 1903 when Burroughs published a tirade against writers who embellished the animal perspective, apparently seeking to dupe the public and disingenuously mixing science and fiction (Lutts, *The*

Nature Fakers, 37). Naturalists objected that the animal writers took liberties such as "describ[ing] the animals' thoughts" (Ibid., 35). Writers such as Seton and Long penned responses to Burroughs' attack, and after holding off for years London himself eventually rebutted: "They believe that man is the only animal capable of reason and that ever does reason. This is a view that makes the twentieth-century scientist smile. It is not modern at all. It is distinctly medieval. President Roosevelt and John Burroughs, in advancing such a view, are homocentric in the same fashion that the scholastics of earlier and darker centuries were homocentric" (Jack London, "The Other Animals," in *No Mentor But Myself: A Collection of Articles, Essays, Reviews, and Letters on Writing and Writers*, ed. Dale L. Walker [Port Washington, NY: Kennikat Press 1979], 110).

11. Jack London, "The Other Animals," 109.

12. Ibid., 109.

13. Zoologist Tristram D. Wyatt writes that "[a]cross the animal kingdom, more interactions are mediated by pheromones than by any other kind of signal" (*Pheromones and Animal Behaviour: Communication by Smell and Taste* [New York: Cambridge University Press, 2006], 4).

14. Temple Grandin and Catherine Johnson, *Animals Make Us Human: Creating the Best Life for Animals* (New York: Mariner, 2010), 74.

15. Vicki Hearne, *Adam's Task: Calling Animals by Name* (New York, Knopf, 1987), 79. This does not mean that canines are non-visual animals. As Temple Grandin notes, "[w]olves communicate heavily through vision" (*Animals Make Us Human: Creating the Best Life for Animals* [New York: Mariner, 2010], 73). Alexandra Horowitz notes that certain dog breeds have "an overabundance of rod photoreceptors in their retinae" and that others have "large, globular eyes for seeing at night" (*Inside of a Dog: What Dogs See, Smell, and Know* [New York: Scribner, 2009], 48). Nonetheless, in addition to sensitive vision Buck and White Fang would also have enormously complex olfactory perception.

16. Lillian Carswell, "The Power of Choice," 302.

17. Charles Darwin, *The Descent of Man, and Selection in Relation to Sex*, ed. James Moore and Adrian Desmond (Toronto: Penguin, 2004), 35.

18. Ibid., 35.

19. Lillian Carswell, "The Power of Choice," 306.

20. Constance Classen, David Howes, and Anthony Synnott, *Aroma: The Cultural History of Smell* (New York: Routledge, 1994), 122.

21. Jack London, *The Call of The Wild, White Fang, & To Build a Fire* (New York: Modern Library, 1998), 119, 121.

22. Ibid., 120 (emphasis added).

23. London, *The Call of the Wild, White Fang, & To Build a Fire*, 40 (emphasis added).

24. This formula—smell is bestial while vision is the terrain of the civilized human—reaches an apotheosis in Sigmund Freud's *Civilization and Its Discontents*. Taking the liberty to construct a narrative of prehistoric humanity, Freud describes a "new way of life, ushered in by man's adoption of an upright gait, against his earlier animal existence." (Sigmund Freud, *Civilization and Its Discontents* [New York: Penguin, 2004], 55).

25. This oft-noted semantic absence has led the poet Dianne Ackerman to call smell "the mute sense." (*A Natural History of the Senses* [New York: Random House, 1990], 6), and this linguistic dearth is not limited to English. Linguist Yanna Popova notes that "smell is characterized by a lack of vocabulary to define it" in "any language" ("'The fool sees with his nose': Metaphoric mappings in the sense of smell in Patrick Süskind's *Perfume*." *Language and Literature*12, no. 2 [2003]: 135–51, 144). Anthropology does offer examples of "osmologies" that challenge Western paradigms (Classen, Howes, and Synnott, *Aroma*, 103). Examples include the Bororo people of Brazil, whose religious beliefs associated gods with particular smells, and the Temiar people of the Malay Peninsula, who believe in what they call the "odour-soul" (Classen, Howes, and Synnott, *Aroma*, 117, 115).

26. Harlan Weaver, "Pit Bull Promises: Inhuman Intimacies and Queer Kinships in an Animal Shelter," *GLQ: A Journal of Gay and Lesbian Studies* 21, no. 2–3 (2015): 343–63, 351.

27. Dana Seitler, *Atavistic Tendencies: The Culture of Science in American Modernity* (Minneapolis: University of Minnesota Press, 2008), 69.

28. Wyatt, *Pheromones and Animal Behaviour*, 274.

29. Irene Tucker, *The Moment of Racial Sight: A History* (Chicago: The University of Chicago Press, 2012), 7.

30. Classen, Howes, and Synnott, *Aroma*, 4.

31. Sigmund Freud, *Civilization and Its Discontents*, trans. David McLintock (Toronto: Penguin, 2004), 55.

32. Sigmund Freud, *Three Case Histories: The 'Wolf Man,' The 'Rat Man,' and The Psychotic Dr. Schreber* (New York: Simon & Schuster, 1996), 198–99.

33. For a compelling and rigorous discussion of Freud's anxious relation to atavism and animality, see Seitler's chapter, "Freud's Menagerie," in *Atavistic Tendencies.*

34. Freud, *Three Case Histories*, 80.

35. Ibid., 80.

36. Havelock Ellis, *Studies in the Psychology of Sex Volume 4* (Philadelphia: F.A. Davis, 1906), 84.

37. Richard Von Krafft-Ebing, *Psychopathia Sexualis: A Medico-Forensic Study*, trans. Harry E. Wedeck (New York: Putnam's, 1965) 58, 61.

38. Ellis, *Studies in the Psychology of Sex Volume 4*, 88–89.

39. Greg Forter, *Murdering Masculinities: Fantasies of Gender and Violence in the American Crime Novel* (New York: New York University Press, 2000), 67.

40. Sigmund Freud, *The Uncanny*, trans. David McLintock (London: Penguin, 2003), 151. Freud writes, specifically: "there is something uncanny about the female genitals."

41. Simone de Beauvoir also notes that psychoanalytic tradition suggests the animalism of the female anatomy, indicated through the profusion of "perverse smells" (*The Second Sex*, trans. Constance Borde and Sheila Malovany-Chevallier [New York: Vintage, 2011], 21.

42. Ellis, *Studies in the Psychology of Sex Volume 4*, 85.

43. Susan Fraiman, "Pussy Panic versus Liking Animals," *Critical Inquiry* 39, no. 1 (2012): 89–115. doi: 10.1086/668051, 100.

44. Nicole Shukin, *Animal Capital: Rendering Life in Biopolitical Times* (Minneapolis: The University of Minnesota Press, 2009), 63.

45. For an informative note about the primal visual scenes of psychoanalysis, see Dana Seitler's *Atavistic Tendencies*, 250n19. Seitler refers most notably to Lacan's mirror stage and the Freudian interpretation of visual signs in dreams.

46. Susan Fraiman, "Pussy Panic versus Liking Animals," 95.

47. Derrida, *The Animal That Therefore I Am*, 4.

48. Jacques Lacan, *The Four Fundamental Concepts of Psychoanalysis*, trans. Alan Sheridan (New York: Penguin, 1994), 73.

49. Grandin, *Animals Make Us Human*, 74.

50. Ibid., 73.

51. Donna Haraway, *When Species Meet* (Minneapolis: University of Minnesota Press, 2008), 21, 226.

52. Susan Fraiman, "Pussy Panic versus Liking Animals,"100.

53. Ibid., 93.

54. Ibid., 100.

55. Derrida, *The Animal That Therefore I Am*, 55.

56. Ibid., 55.

57. Ibid., 32, 33.

58. Jacques Derrida, *Of Grammatology*, trans. Gayatri Chakravorty Spivak (Baltimore: The Johns Hopkins University Press, 1998), 75.

59. André Alexis, *Fifteen Dogs* (Toronto: Coach House, 2015), 15.

60. Marco Caracciolo, "'Three Smells Exist in this World': Literary Fiction and Animal Phenomenology in Italo Svevo's 'Argo and His Master,'' *Modern Fiction Studies* 60, no. 3 (2014): 484–505, 485.

61. Don McKay, *Vis à Vis: Fieldnotes on Poetry and Wilderness* (Wolfville: Gasperau, 2001), 31. Although Alexis cleverly avoids any injunction of scientific realism, he has clearly researched animal behavior and spent time trying to understand dogs. These efforts emerge in his highly readable portrayal of the canine perspective. Though there are weaknesses to his dramatization of dogs—most prominently the frequent, jarring references to "[b]itches in heat" (Alexis, *Fifteen Dogs*, 130)—for the most part Alexis' novel offers a fascinating multi-sensory experience of canine being.

62. The most prominent examples are *Aroma: The Cultural History of Smell* (Routledge, 1994), edited by Constance Classen, David Howes, and Anthony Synnott, and *The Smell Culture Reader* (Berg, 2006), edited by Jim Drobnick, as well as other work cited in this essay by Caracciolo, Oliver, Serres, and Graham.

63. Stephanie Oliver, "Diffuse Connections: Smell and Diasporic Subjectivity in Larissa Lai's *Salt Fish Girl*," *Canadian Literature* 208 (2011): 85–107, 90.

64. Michel Serres, *The Five Senses: A Philosophy of Mingled Bodies*, trans. Margaret Sankey and Peter Cowley (New York: Continuum, 2008), 171.

65. Mark Graham, "Queer Smells: Fragrances of Late Capitalism or Scents of Subversion?" in *The Smell Culture Reader*, ed. Jim Drobnick (New York: Berg, 2006), 315.

66. Ibid., 315.

67. Ibid, 306. Forter also draws compelling connections between olfaction and gay male experience: Forter, *Murdering Masculinities*, 68–9.

WORKS CITED

Ackerman, Dianne. *A Natural History of the Senses.* New York, NY: Random House, 1990.

Alexis, André. *Fifteen Dogs: An Apologue.* Toronto: Coach House, 2015.

Beauvoir, Simone de. *The Second Sex.* Translated by Constance Borde and Sheila Malovany-Chevallier. New York, NY: Vintage, 2011.

Berger, John. "Why Look at Animals?" In *The Animals Reader: The Essential Classic and Contemporary Writings*, edited by Linda Kalof and Amy Fitzgerald, 251–61. New York, NY: Berg, 2007.

Caracciolo, Marco. "'Three Smells Exist in this World': Literary Fiction and Animal Phenomenology in Italo Svevo's 'Argo and His Master.'" *Modern Fiction Studies* 60, no. 3 (2014): 484–505.

Carswell, Lilian. "'The Power of Choice': Darwinian Concepts of Animal Mind in Jack London's Dog Stories." In *America's Darwin: Darwinian Theory and U.S. Literary Culture*, edited by Tina Gianquitto and Lydia Fisher, 302–32. Athens, GA: The University of Georgia Press, 2014.

Classen, Constance, David Howes, and Anthony Synnott. *Aroma: The Cultural History of Smell.* New York, NY: Routledge, 1994.

Darwin, Charles. *The Descent of Man, and Selection in Relation to Sex.* Edited by James Moore and Adrian Desmond. Toronto: Penguin, 2004.

Derrida, Jacques. *Of Grammatology.* Translated by Gayatri Chakravorty Spivak. Baltimore, MD: The Johns Hopkins University Press, 1998.

———. *The Animal That Therefore I Am.* New York: Fordham University Press, 2008.

Ellis, Havelock. *Studies in the Psychology of Sex, Volume 4.* Philadelphia, PA: F.A. Davis, 1906.

Forter, Greg. *Murdering Masculinities: Fantasies of Gender and Violence in the American Crime Novel.* New York, NY: New York University Press, 2000.

Fraiman, Susan. "Pussy Panic versus Liking Animals: Tracking Gender in Animal Studies."*Critical Inquiry* 39, no. 1 (2012): 89–115. doi: 10.1086/668051.

Freud, Sigmund. *Civilization and Its Discontents.* Translated by David McLintock. Toronto: Penguin, 2004.

———. *The Uncanny.* Translated by David McLintock. London: Penguin, 2003.

———. *Three Case Histories: The 'Wolf Man,' The 'Rat Man,' and The Psychotic Doctor Schreber.* New York, NY: Simon & Schuster, 1996.

Graham, Mark. "Queer Smells: Fragrances of Late Capitalism or Scents of Perversion?" In *The Smell Culture Reader*, edited by Jim Drobnick, 305–19. New York, NY: Berg, 2006.

Grandin, Temple, and Catherine Johnson. *Animals Make Us Human: Creating the Best Life for Animals.* New York, NY: Mariner Books, 2010.

Haraway, Donna. *When Species Meet.* Minneapolis, MN: University of Minnesota Press, 2008.

Horowitz, Alexandra. *Inside of a Dog: What Dogs See, Smell, and Know.* New York, NY: Scribner, 2009.

Hearne, Vicki. *Adam's Task: Calling Animals by Name.* New York, NY: Knopf, 1987.

Krafft-Ebing, Richard Von. *Psychopathia Sexualis: A Medico-Forensic Study.* Translated by Harry E. Wedeck. New York, NY: Putnam's, 1965.

Labor, Earle, Robert C. Leitz, III, and I. Milo Shepard. *The Letters of Jack London: Volume One: 1865–1905.* Stanford, CA: Stanford University Press, 1988.

London, Jack. *The Call of the Wild, White Fang, & To Build a Fire.* New York, NY: Modern Library, 1998.

———. "The Other Animals."*No Mentor But Myself: A Collection of Articles, Essays, Reviews, and Letters on Writing and Writers.* Edited by Dale L. Walker. Port Washington, NY: Kennikat Press, 1979. 108–20.

Lacan, Jacques. *The Four Fundamental Concepts of Psychoanalysis.* Translated by Alan Sheridan. New York, NY: Penguin, 1994.

Lutts, Ralph H. *The Nature Fakers: Wildlife, Science, and Sentiment.* Golden, CO: Fulcrum, 1990.

McKay, Don. *Vis à Vis: Fieldnotes on Poetry and Wilderness.* Wolfville: Gaspereau, 2001.

Oliver, Stephanie. "Diffuse Connections: Smell and Diasporic Subjectivity in Larissa Lai's *Salt Fish Girl.*" *Canadian Literature* 208 (2011): 85–107.

Pick, Anat. *Creaturely Poetics: Animality and Vulnerability in Literature and Film.* New York, NY: Columbia University Press, 2011.

Popova, Yanna B. "'The fool sees with his nose': Metaphoric mappings in the sense of smell in Patrick Süskind's *Perfume.*" *Language and Literature* 12, no. 2 (2003): 135–51.

Seitler, Dana. *Atavistic Tendencies: The Culture of Science in American Modernity.* Minneapolis, MN: University of Minnesota Press, 2008.

Serres, Michel. *The Five Senses.* Translated by Margaret Sankey and Peter Cowley. New York, NY: Continuum, 2008.

Shukin, Nicole. *Animal Capital: Rendering Life in Biopolitical Times.* Minneapolis, MN: University of Minnesota Press, 2009.

Tucker, Irene. *The Moment of Racial Sight: A History.* Chicago, IL: The University of Chicago Press, 2012.

Weaver, Harlan. "Pit Bull Promises: Inhuman Intimacies and Queer Kinships in an Animal Shelter." *GLQ: A Journal of Gay and Lesbian Studies* 21, no. 2–3 (2015): 343–63.

Wyatt, Tristram D. *Pheromones and Animal Behaviour: Communication by Smell and Taste.* New York, NY: Cambridge University Press, 2003.

Chapter 8

Do Androids Dream of Derrida's Cat?

The Unregulated Emotion of Animals in Philip K. Dick's Do Androids Dream of Electric Sheep?

Megan E. Cannella

Today, cultural parameters for defining humanity are increasingly intangible and elusive. The human experience is almost endlessly mediated by the pervasive influences of ever-evolving, ever-escalating technological ideologies. The omnipresence of technological mediation ostensibly haunts every aspect of human life in the twenty-first century, leaving animals,[1] who are fundamentally free of this cult of technology, as the only true examples of pure emotion and empathy. I argue that these affects, so often idealized as integral values of humanity, are the guiding force of Philip K. Dick's dystopic, postapocalyptic narrative. In *Do Androids Dream of Electric Sheep?* (1968),[2] animals are the last vestige of affect following the collapse of life on Earth. These animal affects, while often operating at the periphery of Dick's novel, are a constituent part of how the novel mediates the complexities of human and humanoid relations.

In reading the circulation of animal affects in *Do Androids Dream of Electric Sheep?*, this chapter explores how Derrida's insight into the nudity of the animal as an expression of affect in "The Animal That Therefore I Am" enables a reinterpretation of the ways in which the visibility of animals in *Do Androids Dream of Electric Sheep* is manufactured, distorted, and complicated. I argue that the animals of Dick's novel are the final remaining impetus of organic, unregulated emotion in a world of commodified status symbols. Further, I demonstrate how the novel, despite the dominance of the narrative of the protagonist (bounty hunter Rick Deckard), relies on the circulation of animal affects as a guiding and motivating force of action. It is through this pure, unmitigated transmission of animal emotion that the surviving humans of Dick's novel retain any semblance of their preapocalyptic humanity.

In the context of the novel's post-World War Terminus, the decline of humanity is attributed to the inability of humans to express emotion: an ineptitude that is made especially poignant by the unrestrained circulation of animal affects. Set in 2021, Earth is contaminated with postwar radioactive dust and debris, which has caused the extinction of entire species and the migration of humans to Mars. The boundaries between human and nonhuman are irreparably blurred, and the contexts, moralities, and realities of the post-World War Terminus society are similarly obscured. Throughout Dick's novel, the genuine affect produced and elicited by animals both mirrors and mocks the inability of humans, as well as their humanoid counterparts, to participate in any degree of unmediated affect. Addressing the resulting complexities of these fundamental societal obfuscations in his article "Faith and Bad Faith in *Do Androids Dream of Electric Sheep?*" Donald Palumbo asserts:

> Humanity's collective bad faith is also revealed in a trenchant collective amnesia, likewise prompted by guilt [...] And the difficulty humanity has in distinguishing between reality and illusion, between the real and the fake—by far the most common motif in Dick's corpus—is also the most prominent motif in this novel. In addition to contributing to the aesthetic coherence of a story that deals fundamentally with artificial life forms that mimic and are often mistaken for natural life forms, this motif's prominence also reinforces the novel's implicit and persuasive definition of "human."[3]

The bad faith Palumbo identifies as intrinsic to Dick's novel is evidenced in "a trenchant collective amnesia, likewise prompted by guilt." As those remaining on Earth endeavor to reestablish a normative human society, they must suppress their conscious awareness of the warfare that brought the Earth to the point to which it became covered in radioactive debris, a dusty husk of what it used to be. In order to fully accept this new world, which is regularly, artificially mediated by mood organs,[4] the human characters are required to ignore and forget what life was like before they were left behind on Earth, and many of their neighbors recolonized on Mars. To move on with any semblance of emotional stability, they have to ignore the actions that brought them to this place in time, while simultaneously suppressing the guilt they feel for having created the conditions that lead to the World War Terminus.

Given the primacy of grief and guilt that surviving humans on Earth experience throughout the novel, theoretical interpretations of *Do Androids Dream of Electric Sheep?* often focus on the human—non/posthuman dichotomy or on the colonialist aspects of the novel. Often neglected or misunderstood is the role of animals in mediating affective structures. When scholars showcase the representation of Dick's animals, it is notably only in terms of how they preserve or serve humanity. For instance, Nigel Wheale situates animals as pseudo-children[5], while Karyn Ball discusses the power of the "'global animal' as the chiaroscuro surface of human hope and fear: it brims with harbingers of present and future perils, warnings that might be heeded on time if only one could penetrate the still eyes of this anthropomorphic mirror."[6] While both of these critics offer relevant and useful analysis that can be applied to the role of animals in Dick's novel, neither go far enough to endow animals with independence and agency in moderating affect throughout the novel. In my reading of Dick's animal affects, I take my cue from Melissa Gregg, who explains in her book *Cultural Studies' Affective Voices* that "The reason that children have gained such political traction as a site for hegemonic battles is that they share the same function of representing the blank slate, that is, the open and unsullied future that the US has always required as its foundational *raison d'être*."[7] Children and animals are often similarly situated as symbols of hope and humanity's endurance, operating within a larger, human and adult-centric context; however, within this novel, animals are not supporting characters meant to bolster expressions of humanity; rather, animals establish a new hierarchy of emotional expression and affect where they reign supreme over humans.

Due in part to the human need to escape the past and find salvation in the remaining world and culture, animals are exalted to the last true measure of humanity. No longer represented as companions to humans or interpreted as surrogate children or metaphors for human nature, animals in Dick's novel establish and ensure empathy, even in this desolate, post-apocalyptic existence. In fact, animals in this novel largely align with Gregg's understanding of migrants' identity navigation. Gregg writes that "Migrants' particular circumstances encourage a certain competence in negotiating between identities: an originary identity held alongside contemporary hybrid practices while meeting the challenge of forging new relationship of affinity with a foreign culture."[8] Dick bestows the animals in *Do Androids Dream of Electric Sheep?* with a unique capacity to perpetuate affect: they are the guardians of unmediated emotion in a postwar world.

It is important to recognize that any human expression of affect mediated either through technology or social expectation cannot be considered authentic compared to animal expressions of emotion, which are both independent and unmediated. Wisam Kh. Abdul-Jabbar argues, for instance, that animal mimicry "mocks, firstly, humans' vanity as humans fancy they can care for actual life and, secondly, because the ability or the luxury to own an animal has become a symbol of status and not a sign of animal care. This kind of deception corresponds to the deceptive ways of the colonizers to justify their colonial campaign in the name of reforming the Other."[9] Similar to Palumbo's discussion about the bad faith of this community, Abdul-Jabbar's exploration of mimicry in the novel begins with a conversation about the role of animals in the society that Dick creates. In the context that Dick provides readers, animals, regardless of whether or not they are a product of artificial manufacturing, serve to maintain and demonstrate humanity through their unregulated expressions of affect. Whereas humans regulate emotions with mood organs and androids are incapable of empathy, animals are the only entity in this novel regularly engaging in, and therefore perpetuating, unmediated affect. In this novel, animals are exclusively situated to embody and exude affective nudity.

Stationed on a war-torn, radiation-saturated planet, the surviving humans are faced with feelings of inferiority and purposelessness. Yet the animals, android or not, all require care and attention, motivating humans to go to work and continue being productive in an otherwise destroyed and hopeless society. As long as animals exist, humans can maintain the allusion of species superiority and hierarchy through exertions of ownership. Dick extrapolates upon this hierarchy though discussion of human emigration from Earth to Mars, explaining, "A meager colonization program [of Mars] had been underway before the war, but now that the sun had ceased to shine on Earth, the colonization entered an entirely new phase. [...] Under U.N. law each emigrant automatically received possession of an android subtype of his choice"[10] The humans deemed worthy of emigration are also deemed as quality colonizers who automatically receive android servants upon arriving to Mars. Were it not for the existing hierarchy where humans on Mars are superior to those left behind on Earth, androids escaping from slavery on Mars would not need to be hunted. Animals are kept, and androids are hunted because the humans left on Earth insist on invoking an innate kind of ownership and entitlement to both the planet and any decidedly nonhuman entity that they deem as "Other"—thus, keeping their own othering at bay. If it were not for the colonial power that humans continue to derive from animal ownership, it would be much easier for them to stop viewing androids as the Other, allowing androids to take over, or at least share, what is left of the Earth.

Through its complicated and, at times, convoluted treatment of animals, *Do Androids Dream of Electric Sheep?* exemplifies the complexities of nakedness and visibility that Derrida puts forward in "The Animal That Therefore I Am." In the postapocalyptic world that Dick creates, animals are the only remaining genesis of true, genuine emotion, existing free of regulation, unlike their human and humanoid counterparts. This positions animals as the sole arbiters of unmediated affect, thus situating animals as superior in this postapocalyptic hierarchy of affective humanity. Dick's narrative treatment of animals mirrors Derrida's assertion that, "The animal, therefore, is not naked because it is naked. It doesn't feel its own nudity. There is no nudity 'in nature.' There is only the sentiment, the affect, the (conscious or unconscious) experience of existing in nakedness. Because it is naked, without existing in nakedness, the animal neither feels nor sees itself naked."[11] The only trace of "nature" or "nakedness" in Dick's novel resides with the animals. Even though the citizens of Earth are now able to regulate their emotions with the mood organ, animals elicit emotions that transcend mediated affect. When an animal dies, or if there is not enough money to afford an animal in the first place, or if a former lover chooses his animal above all else, raw emotion is uncontrollably, inevitably exposed, even in androids that are designed to, at best, only simulate empathy. By putting *Do Androids Dream of Electric Sheep?* in conversation with Derrida's assertions about animal nakedness and affect, the animals in this novel can be repositioned as the focal point of the novel. In a dystopic novel, where the human and android characters clamor to express true affect—often manifested as searches for love, attention, connection, and/or dominance—authenticity is only afforded to the animals. By way of the continually mediated and regulated lives that the humans endure, the animals are assigned full emotional jurisdiction in this novel.

Despite the shame initially felt by the protagonist of the novel, Rick, due to his inability to purchase a nonandroid, organic sheep, he is nevertheless motivated to make his bounties, collect his reward money, and obtain an animal as a status symbol. This status symbol, as William S. Haney explains in his discussion of the effect of limited human development in Dick's novel, is attributed to the animal as "a sign of empathy," regardless of "whether the animals are real or electric."[12] While Rick and his wife did not have enough money to escape the destroyed, radioactive Earth, having a nonandroid pet would help maintain what social collateral they have left. We learn early in the novel, for example, that "Loitering on Earth potentially meant finding oneself abruptly classed as biologically unacceptable, a menace to the pristine heredity of the race. Once pegged as special, a citizen, even if accepting sterilization, dropped out of history."[13] Even if he is now deemed biologically unacceptable for Mars, Rick wants to maintain at least the allusion of the highest class of life achievable on Earth, and owning a nonandroid animal is the quickest way to perpetuate such an allusion. I argue that virtually everything that takes place in the novel—most notably Rick's strained marriage, his bounty hunting adventures, and his affair with an android—is secondary to Rick's quest for authentic animal ownership, a fact that ultimately supports Derrida's assertions on animal visibility. Animal visibility thus becomes paramount throughout the novel because it becomes the motivating force for all main characters.

As Rick plots and schemes to earn enough money to purchase a large, nonandroid, organic animal as a way to increase his social status, his wife, Iran, dedicates herself to maintaining their android animals in an attempt please Rick and salvage their marriage. As Rick, Iran, and many other characters in this novel are overwhelmed by the challenge of assimilating to life after World War Terminus and the subsequent changes in societal mores, the animals are oblivious to these changes. All animals are outside of, and free of, the societal norms humans have imposed first upon Earth and then upon

Mars. As Iran dutifully schedules her moods, she purposefully includes "a six-hour self-accusatory depression."[14] She is aware that scheduling despondent moods is the only way she can experience a range or depth of emotions. Meanwhile, animals have no such need to regulate: they are able to exist and form attachments with owners without conscious effort. The continued juxtaposition between mediated and unmediated lives, showcased by Dick through the human-animal binary, exemplifies what Derrida describes as the unconscious affect of animal visibility. He explains that, "On the other hand, because the animal is naked without consciousness of being naked, it is thought that modesty remains as foreign to it as does immodesty. As does the knowledge of self that is involved in that."[15] This lack of consciousness even extends to the android animals that are incapable of realizing that they are operating within a primarily organic nonandroid reality, where they are simply products that have been produced to fill a void for consumers.

The emotions incited by this consumerist void are demonstrated by Rick's desperate pursuit of animal ownership and its associated social status throughout the novel. When Rick is talking to his neighbor about the neighbor's pregnant horse, the narrator reveals Rick's inner angst over not owning a nonandroid, organic animal of his own: "He wished to god he had a horse, in fact any animal. Owning and maintaining a fraud had a way of gradually demoralizing on. And yet from a social standpoint it had to be done, given the absence of the real article. He had therefore no choice except to continue."[16] Rick goes on to explain, perhaps to both comfort and validate himself, that the relationship and responsibility that he had with his android sheep is almost identical to that which one would have with a nonandroid, organic animal:

> You feel the same [caring for the android animal]; you have to keep your eye on it exactly as you did when it was really alive. Because they break down and then everyone in the building knows. I've had it at the repair shop six times, mostly little malfunctions, but if anyone saw them—for instance one time the voice tape broke or anyhow got fouled and it wouldn't stop baaing—they'd recognize it as a *mechanical* breakdown. [. . .] The repair outfit's truck is of course marked 'animal hospital or something.' And the driver dresses like a vet, completely in white.[17]

In this scene, Dick explores the impact of representing "authentic" animality in a society preoccupied with the ownership and care of an android animal. While Rick's need to obtain a nonandroid, organic animal can be dismissed as one man's obsession, the societal collateral that animals hold is made apparent in the elaborate veterinarian façades that android animal repair services assume. Even android animals are provided with veterinarians, fostering an image of authenticity by disguising the crassness of technicians in a repair warehouse. As this novel thrives upon this kind of deception, it leaves the reader, as well as the characters, wondering what, if anything, is genuine, and would we be able to recognize the genuine, even if we were to encounter it? On the post-World War Terminus Earth, nonandroid, organic animals are one of the last semblances of authenticity because "[e]mpathy, evidently, existed only within the human community, whereas intelligence to some degree could be found through every phylum and order including arachnida."[18] Still, despite the fundamental exclusivity of owning a nonandroid, organic animal, Rick's vanity and greed cannot be overlooked. While he struggles to earn enough to purchase a nonandroid, organic animal, he continues to dream extravagantly, as he confesses to his neighbor: "I don't want a domestic pet. I want what I originally had, a large animal. A sheep, or if I can get the money, a cow or a steer or what you have, a horse."[19] The narrator elaborates on Rick's reasoning and aspirations further, stating:

> The bounty from retiring five andys[20] would do it, [Rick] realized. A thousand dollars apiece, over and above [Rick's] salary. Even if the listing in Sidney's Animal and Fowl is in italics. Five thousand dollars—but, he thought, the five andys first have to make their way to Earth from one of the colony planets; I can't control that, I can't make five of them come here, and even if I could, there are other bounty hunters with other police agencies thought the world. The andys would specifically have to take up residence in Northern California, and the senior bounty hunter in this area, Dave Holden, would have to die or retire.[21]

Here, Rick is concerned about earning the money to afford the large animal he desires, but he also entertains ideas of the perfectly choreographed reality that would need to be in place for him to accomplish his goal. This desire can be read as a type of neurosis, similar to what Donna Haraway identifies in *The Companion Species Manifesto*. Haraway asserts that tools, "such as domestic animals (dogs) and computers (cyborgs), [are] evidence of a neurosis that I call humanist technophiliac narcissism" and can be contrasted with "the superficially opposed idea that dogs respite human beings' souls by their unconditional love," which "might be the neurosis of caninophiliac narcissism."[22] In *Do Androids Dream of Electric Sheep?*, Rick is blighted by a variation of caninophiliac narcissism. Still, Rick's desire and desperation for a nonandroid, organic animal is grounded in his determination to usurp the kind of unregulated, unmediated power that animals possess. Accessing such power via animal companionship or even caninophiliac narcissism is not sufficient. Rather, Rick craves dominion over the type of power and existence that animals in this world have, though he perpetually fails to achieve such status.

As such, animals represent the innocence and authenticity of the pre-World War Terminus society, the memory of which seems to be constantly hovering at the periphery as a shameful reminder of the devastation caused by humans. As humans struggle on a contaminated Earth and endeavor to colonize Mars, android servants attempt to escape colonization by fleeing to the contaminated Earth where they can assume the roles of humans. As the reader follows this narrative, it is hard not to compare this to the world as we know it today—pre-World War Terminus. The dwellings are the same, and many of the day-to-day practicalities of this postapocalyptic reality mirror contemporary life in the Western world, or as it may reasonably be in a few years. Just as those left on the postwar Earth must engage in forced amnesia and ignorance about their role in creating this postapocalyptic society, the reader is continually forced to confront the remnants of such destructive human behavior in order to grapple with the affects of shame that Dick's characters are seemingly unequipped to engage. Derrida addresses this inability to cope with guilt and responsibility, stating that, "I have trouble repressing a reflex of shame. Trouble keeping silent within me a protest against indecency. [...] The gaze of a seer, a visionary or extra-lucid blind one. It is as if I were ashamed, therefore, naked in front of this cat, but also ashamed of being ashamed."[23] In Rick's efforts to afford an animal, he is constantly confronted by his own failures—failure to be smart enough to emigrate to Mars, failure to find genuine romantic love with either his wife or his mistress, failure to be professionally successful enough to afford the level of fine living he craves.

As Rick negotiates his efforts and desires to purchase a large, nonandroid, organic animal, his shortcomings are always at the forefront, positioning him as vulnerable and prostrated in comparison to his mythical future animal. It is the presence and predominance of animals in this novel that establishes a continued sense of innocence and mobility in the contaminated, stagnant world through which Rick must navigate and hunt. In a world where android animals need as much care and attention as organic animals, there must also be veterinary clinics, such as the Van Ness Pet Hospital, for the android animals, which authentically come complete with disease circuits. When Isidore, an employee at the Van Ness Pet Hospital and a chickenhead,[24] picks up the ailing cat of Mrs. Pilsen, the reader gains appreciation for just how life-like the android animals are. Unable to charge the cat or revitalize it in transport, Isidore simply delivers the cat to the technicians at the hospital to be properly repaired and returned to its owner. It is only upon arriving at Van Ness Pet Hospital with Mrs. Pilsen's cat that Isidore and the reader realize that the cat is not an android. Hannibal Sloat, Isidore's boss, is troubled by this mix-up of bringing an organic animal to this android animal repair facility, stating, "But it's the waste that gets me. The loss of one more living creature. Couldn't you tell, Isidore? Didn't you *notice* the difference?"[25]This scene speaks to the dire depletion of living creatures and to Isidore's lost ability to negotiate the intricacies of humanity. Another employee at the repair facility helps to explain this both to Sloat and the audience:

> "I don't think Isidore can tell the difference," Milt said mildly. "To him they're all alive, false animals included. He probably tried to save it. [...] It probably was so far gone it wouldn't have made it anyhow [...] Let the chickenhead off the hook, Han. He's got a point; the fakes are beginning to be darn near real, what with those disease circuits they're building into the new ones. And living animals do die; that's one of the risks in owning them. We're just not used to it because all we see are fakes."[26]

The reality of the false animals obscures the ways in which diseased animals are treated and also falsifies human interaction with these false, android animals.

These manufactured animals are not simply objects for one to own. As Derrida suggests, "But even before that identification, it comes to me as this irreplaceable living being that one day enters my space, into this place where it can encounter me, see me, even see me naked."[27] Animals in this novel, whether android or not, enter personal space, encountering their owners and more importantly, seeing their owners through an unmediated lens. Dick's positioning of animals as unmediated, affective voyeurs, especially in respect to their owners, facilitates the reading of this novel in terms of Derridian nudity. To this end, Derrida explains, "at that strange moment when, before the event, before even wanting it or knowing it myself, I am passively presented to it as naked, I am seen and seen naked, before even seeing myself seen by a cat."[28] Since animals are a status symbol in this postapocalyptic society, animals establish human identity simply by virtue of their existence. It is Rick's ownership of a nonandroid, organic animal that determines not only his larger societal worth but also his own sense of self-worth. Rick has been able to outwardly pretend that his android animal was in fact organic, but that allusion (or perhaps delusion) dissolves when he is face to face with either his android sheep or eventually his nonandroid, organic goat, because as Derrida posits, "Before even seeing myself or knowing myself seen naked. I am presented to it before even introducing myself. Nudity is nothing other than that passivity, the involuntary exhibition of the self. Nudity gets stripped to bare necessity only in that frontal exhibition, in the face-to-face."[29] This society is constructed in such a way that much is obscured and manipulated; however, animals continue to exclusively retain access to unmediated affect.

Human owners encounter their android, electric animals without regulation, just as they would a nonandroid, organic animal, creating a space of vulnerability only available in the presence of an animal. Unlike the backflipping robotic dog or the Tamagotchis[30] that were attached to the key rings and backpacks of many children in the 1990s, android animals of Dick's novel are not objects of play soon discarded in a drawer or closet. Unlike a Tamagotchi, an android animal that is not properly cared for will not reboot at the end of the day, nor will it automatically generate new lives. Animals, android or not, require the care and attention of their owner in a genuine, unscripted way, that cannot be scheduled or regulated as the other emotions humans experience in the novel. Just like an organic animal, android animals require regular visits to places like the Van Ness Pet Hospital, and when they do not go for regular checkups, they will eventually deteriorate.

For instance, upon learning that her pet cat has died, Mrs. Pilsen's exploration of replacement options center around sparing her husband from the grief of this loss. She ponders, "Maybe I ought to commission an electric replacement of Horace [her cat] but without Ed [her husband] ever knowing; could it be so faithful a reproduction that my husband wouldn't be able to tell?"[31] In response, she is told by the repair facility that electric replacements are generally only convincing to the casual passerby or neighbor. An electric replacement should not be able to fool an engaged and committed owner because, as Haraway explains, "the major demand on the human is precisely what most of us don't even know we don't know how to do—to wit, how to see who the [animals] are and hear what they are telling us, not in bloodless abstraction, but in one-on-one relationship, in otherness-in-connection."[32] Because of the presumably intimate connection between pet and owner, it is assumed that an owner will always be able to tell when his or her pet had been replaced by an electric reproduction. Regardless, Mrs. Pilsen assures the repair technician, and likely herself, that "Ed never got physically close to Horace, even though he loved him; I was the one who took care of all Horace's personal needs such as his sandbox. I think I would like to try a false animal, and if it didn't work then you could find us a real cat to replace Horace. I just don't want my husband to know; I don't think he could live through it. That's why he never got close to Horace; he was afraid to."[33]Although Mrs. Pilsen is deliberately duping her husband in a fundamentally disturbing way, her crisis of conscious over whether or not to replace Horace with an electric reproduction is emblematic of the faulty empathy that humans perpetuate among themselves.

Mrs. Pilsen's husband, despite his unwillingness to get too close to Horace, is still hopelessly connected to this cat. His reluctance to engage with Horace for fear of experiencing an unregulated emotional response elucidates how, as Derrida argues, "The experience of the seeing animal, of the animal that looks at them, has not been taken into account in the philosophical or theoretical architecture of their discourse. In sum, they have denied it as much as misunderstood it."[34] In a postapocalyptic world devoid of any significant hope or opportunity, humans, like Rick, crave animal ownership as a means to a more lucrative social status and to assuage humanity's guilt over World War Terminus. However, the affective disconnect between animals and humans prevents animal ownership or animal interaction from eliminating any of the affective mediation that regulates this world. This inability to understand and meaningfully recognize the animal is certainly the case for the Pilsens. By replacing Horace with an electric reproduction, Mrs. Pilsen is creating more distance between not only the replacement cat and her husband, but also, inadvertently, between her husband and herself.

The duplicitous reproduction of Horace is emblematic of the problematic deconstruction of the human ability to empathize. It strips away the truth or nakedness on which Derrida reflects, as the interactions between new-Horace and Mr. Pilsen are now mediated through Mrs. Pilsen's deception. Unbeknownst to Mr. Pilsen, the nakedness and truth that once existed between himself and Horace has been excised. While Horace can be replicated, his capacity for genuine affect cannot. Even though Mrs. Pilsen's actions are supposedly motivated by her desire to spare her husband from the pain of grief, she is really just creating an alternate reality, the maintenance of which depletes the remaining humanity and/or empathy that exists in her home. Much like Rick, whose job as a bounty hunter is to determine what is and is not an android based on a test devised to detect empathy, Mrs. Pilsen's deception results in the fragmentation of human empathy, leaving animals as the only true expression and impetus of empathy left in this society.

Mrs. Pilsen's attempt to preserve what she deems as an integrally "human" affective relationship with their cat demonstrates the gendered dimension of animal affects. In her discussion of the human-android relations of Dick's novel, N. Katherine Hayles argues that "Occasionally male characters, who are constantly in danger of being sucked into the maelstrom of conflicting impulses that the schizoid android evokes, take revenge on the woman who attracts them, presumably with authorial sanction."[35] Hayles's assertion of a contemptuously gendered relationship between human and android supports the fact that the status of the animals as the truest evidence of empathy is seen most clearly through the main female characters in this novel, Iran and Rachael, who equally demonstrate the supremacy of animals in a society where empathy is rarified and commodified. While Rick works tirelessly with the constant hope of purchasing a nonandroid animal, his drive and ambition does not apply to addressing his tumultuous relationships with the women—both android and nonandroid—in his life. These heteronormative relationships are complicated by the mediation of affect inherent to this society, with Rick's feelings for the women always being artificially reciprocated.

When he starts to develop feelings for Rachael, an android, she assures him that if they have sex, he will not be able to retire her. He, perhaps consumed by his own humanity, which fundamentally cannot comprehensively or productively understand the stark difference between humans and androids, still pursues intimacy with Rachael, apparently in the hope that love or infatuation or lust can somehow transcend her nonhuman sensibilities. He seems to hold this intimacy and affection for Rachael, even though, mere moments before Rick and Rachael sleep together, she is reflecting on the fact that Rick must retire an android that looks exactly like her. Rachael worries that Rick will mistakenly retire her, and sadly reflects, "We are machines, stamped out like bottle caps. It's an illusion that I—I personally—really exist I'm just representative of a type."[36] Such expressions of empathy are not within an android's capacity, and this admission of kindred connection and identification seems to bolster Rick's hope, and perhaps the reader's expectation as well, that Rachael and Rick will somehow realize their fundamentally doomed attraction. This hope is escalated even further when Rachael and Rick are having an intimate post-coital conversation, despite the fact that it is far more likely that Rachael is just performing the role of Rick's romantic interest.

Rachael interrogates Rick about his wife and his true feelings for Rachael, thus fulfilling her role in Rick's life as the archetypical mistress.[37] In this moment, as their post-coital cocoon begins to dissolve, Rick, Rachael, and the voyeuristic reader are immediately reminded that this cross-species love affair cannot be sustained. As the elation begins to dissipate, Rachael suggests that she and Rick could live together, despite him being married, if only she were alive. It is at this point that Rick explicates the inherent fragility of any potential relationship with Rachael, explaining that, "'Legally you're not [alive]. But really you are. Biologically. You're not made out of transistorized circuits like a false animal; you're an organic entity.' And in two years, he thought, you'll wear out and die. Because we never solved the problem of cell replacement, as you pointed out. So I guess it doesn't matter anyhow."[38] Here, as is the case with many illicit affairs, the fragility and temporality of the encounter cannot be pushed off or avoided for as long as the participants want to or need it to be. Rachael, biologically akin to the android animals Rick desires, is a temporary placeholder in Rick's quest toward animal ownership. She offers him an outlet for his emotion, in a decidedly unregulated and culturally unapproved way, and while Rick basks in the afterglow of this affective fix, he must continue on to achieve his larger goals of nonandroid, organic animal ownership.

The temporal fragility of Rick and Rachael's affair is reinforced through the attention Rick pays to his newly acquired goat. When Rachael first invites Rick to come to bed with her, he rebuffs her because of his allegiance to his goat, responding, "I bought a black Nubian goat [...] I have to retire the three more andys. I have to finish up my job and go home to my wife."[39] Before mentioning even his wife or his job, he refers to his newly purchased, nonandroid goat. Once their post-coital bliss is fully dissolved, Rachael berates Rick for looking down on her for being an android but also being weak like the bounty hunters she had slept with before. She accuses him of being too weak to retire any other androids after having sex with her and dismisses him to go home to take care of his goat. She continues this cold, and somewhat triumphant dismissal by asserting, "'That goat [...] You love that goat more than me. More than you love your wife, probably. First the goat, then your wife, then last of all—' She laughed merrily. 'What can you do but laugh?'"[40] Here, even if we set aside Rachael's tone as that of a seemingly jilted lover, Rachael's assessment of Rick's goat-centric priorities is important. As Hayles observes, "It hints at the ironic fact that humans revere animal life but feel free to kill intelligent android life, and it also suggests that Rachael, although she uses Deckard for her own political purposes, cares about whether he cares for her."[41] Rick's blind prioritization of animal life over android life mimics the nudity of Derrida's cat. Rick is no more cognizant of his fetishization of animal ownership than Derrida's cat is of its nudity. Rick's exultation of the animal is intrinsic to his being, similar to the way in which nakedness is inherent to the cat. As Rick continues on his quest to retire the last three andys that he is hunting, he distances himself even further from the emotion that has manifested between him and Rachael. He is propelled naturally and instinctively to do what he must in order to protect and care for his new goat, even at the expense of his lover, a truth which is ultimately the impetus for Rachael to execute Rick's goat by pushing it off the roof of Rick's building to its death.

When Rachael murders Rick's goat, the empathy the goat embodies and instigates becomes emblematic of the Derridian nudity, as animalistic, natural productions of empathy peak in the novel. Until this point in the novel, it has been made abundantly clear that androids are not capable of empathy. In this scene, the empathy Dick has imbued animals with becomes all powerful—influencing even those previously immune and impervious to affective displays. Rick rejecting Rachael in favor of an animal capable of empathetic affect enables Rachael to transcend her programming. In this moment of confrontation, Rachael engages in a completely humanistic, emotional display of revenge. When her lover supplants her to virtually everything else in his life, Rachael lashes out. The fetishized bond between Rick and his goat imbues Rachel with the ability to, uncharacteristically, participate a web of empathy and revenge. As a result, Rachael's pain is reified and validated when she realizes that Rick experiences her pain. Yet it is only through his connection to his goat that Rick is able to feel freely. Unable to share his emotions with his wife or lover, Rick is deeply impacted by the murder of his goat, a loss that extends to the grief he feels as his marriage disintegrates.

His wife Iran's concerns over the murder of the goat are compounded when Rick brings home an android toad, which he believes to be a nonandroid, organic, extinct toad. While human-androids are generally aware of their lack of empathy and the fact that they are actually androids, android animals have no such cognizance. Much like their nonandroid counterparts, the android animals are "naked without the consciousness of being naked." Despite knowing that these similarities exist between nonandroid and android animals, Iran is still hesitant to reveal the toad's hidden electrical panel to Rick. Once exposed, it seems as if Rick and the audience are prompted to wonder how Rick could have ever missed it in the first place. Regardless, even electric animals still must be fed and cared for, and Rick sadly admits, "The electric things have their lives, too. Paltry as those lives are."[42] As Rick tries to fall asleep at the end of the novel, dealing with the physical and emotional drain of his actions and losses over the course of the novel, Iran, sees to the care of the toad:

> Opening the phone book, she looked in the yellow pages under animal accessories, electric; she dialed and when the saleswoman answered, said, "I'd like to order one pound of artificial flies that really fly around and buzz, please." [. . .] The clerk said, "For a toad I'd suggest also a perpetually renewing puddle, unless it's a horned toad, in which case there's a kit containing sand, multicolored pebbles, and bits of organic debris. And if you're going to be putting it through its feed cycle regularly, I suggest you let our service deportment make a periodic tongue adjustment. In a toad that's vital."
> "Fine," Iran said. "I want it to work perfectly. My husband is devoted to it."[43]

This scene at the end of the novel is immensely important to understanding the overall role of animals to the narrative because in it Rick secures an animal not so much through his own actions but by way of his wife's extemporaneous displays of empathy. Although Rick does not have the nonandroid, organic animal he has longed for, his wife's efforts to create and maintain the allusion of such authenticity connect them to one another in a new way.

After Rick has defeated the androids he was hunting and grapples with his own sense of identity and humanity during his affair with Rachael, his desperate need for a nonandroid, organic animal climaxes. Iran, seeing Rick's pain and desperation, yet unable to alleviate his grief through any intimate human-to-human connection, endeavors to ensure that the toad her husband brings home from the Oregon desert is as lifelike as possible. By purchasing everything that the electric toad needs to thrive, Iran acknowledges the real, unconditional love that Rick seeks from the frog, and, perhaps rightly, does her best to supplant the reality that "[a]ndroids are human constructs whose interiority is a fabrication, not a genuine inner experience witnessed by pure consciousness or the internal observer, which they lack."[44] Throughout the novel, Rick invests a substantial amount of his identity and esteem in his relationship with androids and always expects to eventually purchase a nonandroid, organic animal. In the end, due to his inflated expectations, for Rick to realize the degree to which he has poured himself into empty vessels would be devastating. As Haney continues, "Androids lack the basis of volition because they have no conscious awareness. We can ascribe volition, aesthetically speaking, to them imaginatively; however, they lack the free will of human characters."[45] If we assert that androids and animals have no capacity for conscious awareness, then throughout the course of the novel, Rick has been consumed emotionally and mentally by subjects that can never truly reciprocate. Rick continually exerts humanity and is never reciprocally fulfilled. Just as Derrida's cat is unable to acknowledge its nakedness or in fact even be naked due to the fact that it does not recognize or comprehend the concept of nakedness, Rick, by the end of the novel, is unable to address or recognize his humanity, for at this point, he is no longer operating within a paradigm of humanity that situates androids in a purely utilitarian space.

Viewing the animals in Dick's *Do Androids Dream of Electric Sheep?* through a Derridian lens complicates the perceived message and agenda of this novel by privileging the animal narrative as the only true expression of empathy and thereby supplanting humans within the cultural hierarchy of the novel. Though battles within the human-machine binary, as well as the human-humanity binary, lead to popular understandings of this novel, understanding the role of empathy as it applies to the animals and how humans interact with these pure forms of empathy in this novel is more relevant for twenty-first-century audiences. As humans are, now more than ever, investing their humanity in machines and technology, they lose the ability to ever attempt the nakedness of Derrida's cat. As we establish web presences and social media personas, the true self becomes increasingly elusive, and just like the mood organ that Rick and his wife uses, the hashtag and emojis are used to, if not impose fabricated emotions, at least manufacture perceived emotions.

In today's world, the collaboration between human and technology is becoming increasingly seamless. In large part, we have given up much of our privacy and propriety for the manufactured community and convenience that evolving technologies afford us. We continue to engage with popular culture that positions humans and technology as peers, despite how off-putting, problematic, or horrifying, and we maintain human interaction with domesticated animals as effusively emotional, allowing the audience to use these animals as vehicles to understand humanity or the human condition. However, what Dick accomplishes with animals in this novel transcends the idea that animals are merely subservient to humans and the concept of humanity. He allows animals their true autonomy and purity, highlighting the fraught and obscured nature of humanity.

Rick is unable to achieve his goals, as they are grounded in allusion, which itself is grounded in the life Rick knew before World War Terminus. Given the mental and physical impacts that exist as side effects of this war, it is not hard to conceive that Rick's perception, as well as that of those around him, is tremendously flawed. Rick's attempts to retain a version of his prewar life are painfully futile. Rick's innate futility is only clearly seen through the purity of animals in this novel. Animals remain free from the agenda and nostalgia that persists in crippling the potential of humanity, in order to meaningfully and somewhat successfully establish traces of prewar humanity in this new world. Imbued with the kind of Derridian nakedness that fundamentally eludes Rick, and all human and humanoid characters in this novel, Dick's *Do Androids Dream of Electric Sheep?* situates animals as the one remaining legacy of humanity and affect in the desolate, dystopic reality of this narrative.

NOTES

1. For the purposes of this paper, the term "animal(s)" will refer to both electric, android animals and nonandroid, organic animals, unless otherwise specified.

2. Many are familiar with this book because it is the inspiration for the film *Blade Runner* (1982), and it is commonly thought of a sci-fi thriller where bounty hunters pursue androids, and the lines between human and nonhuman become blurred.

3. Donald Palumbo, "Faith and Bad Faith in *Do Androids Dream of Electric Sheep?*" *The Journal of Popular Culture* 46.6 (2013): 1284, 1286–87.

4. A mood organ is a device used to program and regulate mood. An example of its problematic nature is best witnessed in the scene where Iran reveals to Rick that her "[mood] schedule for today lists a six-hour self-accusatory depression." Philip K. Dick, *Do Androids Dream of Electric Sheep?* (New York: Del Rey, 1968, 4). Rick responds to this revelation by explaining, "It defeated the whole purpose of the organ [to schedule oneself for depression]." Dick, *Androids*, 9. This raises the question of how important it is to maintain a traditionally human range of emotion versus just succumbing to the artificial contentedness facilitated by this mood organ.

5. Nigel Wheale, "Recognising a 'human-Thing': Cyborgs, Robots and Replicants in Philip K. Dick's *Do Androids Dream of Electric Sheep?* and Ridley Scott's *Blade Runner*" *Critical Survey* 3.3 (1991): 298.

6. Karyn Ball, "Primal Revenge and Other Anthropomorphic Projections for Literary History," *New Literary History*, 39.3 (2008): 536.

7. Melissa Gregg, *Cultural Studies' Affective Voices* (New York, NY: Palgrave, 2006), 96.

8. Ibid., 75.

9. Wisam Kh. Abdul Jabbar, "Colonial Mimicry in Philip K. Dick's *Do Androids Dream of Electric Sheep?*" *Inquire: Journal of Comparative Literature* 2.2 (2012): np.

10. Dick, *Androids*, 16.

11. Jacques Derrida, "The Animal That Therefore I Am (More to Follow)" in *The Animal That Therefore I Am*, ed. Marie-Louise Mallet, trans. David Wills. (New York: Fordham University Press, 2008), 5.

12. William S. Haney, "Philip K. Dick's We Can Build You and *Do Androids Dream of Electric Sheep?* The Effect of Limited Human Development," *Consciousness, Literature and the Arts* 10.1 (2009): np.

13. Dick, *Androids,* 16.
14. Ibid., 4.
15. Derrida, "The Animal," 5.
16. Dick, *Androids*, 9.
17. Ibid., 12–13.
18. Ibid., 30–31.
19. Ibid., 14.
20. Andys are androids that served as servants to the humans who colonized Mars. Andys escape to Earth in an attempt to free themselves from their lives of servitude. Once on earth they are hunted and exterminated, colloquially referred to as "retired," as the androids are easily mistaken for humans, aside from the android's inability to experience empathy.
21. Ibid.
22. Donna Haraway, *The Companion Species Manifesto* (Chicago: Prickly Paradigm Press, 2003), 33.
23. Derrida, "The Animal," 4.
24. Those humans left on earth who have been so affected by the radioactivity in World War Terminus that they are severely mentally impaired and unable to fully function in society. Isidore stands up against his employer and coworker, explaining that he does not want to be called a chickenhead, as the radioactive dust has effected them as well—with Sloat "damn near blind and in a couple of years [...] won't be able to hear" and Sloat describing Milt, Isidore's coworker as having skin "the color of dog manure." Dick, *Androids*, 79.
25. Ibid., 77.
26. Ibid, 77–78.
27. Derrida, "The Animal," 9.
28. Ibid., 11.
29. Ibid.
30. A virtual reality pet that was a hugely popular toy/fad in the 1990s.
31. Dick, *Android*s, 81.
32. Haraway, *Companion Species*, 45.
33. Dick, *Androids*, 81–82.
34. Derrida, "The Animal," 14.
35. N. Katherine Hayles, *How We Became Posthuman* (Chicago: University of Chicago Press, 1999), 178.
36. Dick, *Androids*, 189.
37. Ibid., 197.
38. Ibid.
39. Ibid., 191.
40. Ibid., 202.
41. Hayles, *Posthuman*, 173.
42. Dick, *Androids*, 241.
43. Ibid, 243–44.
44. Haney, "Limited Human Development," np.
45. Ibid.

WORKS CITED

Abdul Jabbar, Wisam Kh. "Colonial Mimicry in Philip K. Dick's *Do Android's Dream of Electric Sheep?*" *Inquire: Journal of Comparative Literature* 2.2 (2012): np.

Ball, Karyn. "Primal Revenge and Other Anthropomorphic Projections for Literary History." *New Literary History* 39.3 (2008): 533–563.

Derrida, Jacques. "The Animal That Therefore I Am (More to Follow)." In *The Animal That Therefore I Am*, edited by Marie-Louise Mallet, translated by David Wills, 1–51. New York, NY: Fordham University Press, 2008.

Dick, Philip K. *Do Androids Dream of Electric Sheep*? New York, NY: Del Ray, 1968.

Gregg, Melissa. *Cultural Studies' Affective Voices*. New York, NY: Palgrave, 2006.

Haney, William S. "Philip K. Dick's We Can Build You and *Do Androids Dream of Electric Sheep?* The Effect of Limited Human Development." *Conciousness, Literature and the Arts* 10.1 (2009): np.

Haraway, Donna. *The Companion Species Manifesto*. Chicago, IL: Prickly Paradigm Press, 2003.

Hayles, N. Katherine. *How We Became Posthuman*. Chicago, IL: University of Chicago Press, 1999.

Palumbo, Donald. "Faith and Bad Faith in *Do Androids Dream of Electric Sheep?*"*The Journal of Popular Culture* 46.6 (2013): 1276–1288.

Wheale, Nigel. "Recognising a 'human-Thing': Cyborgs, Robots and Replicants in Philip K. Dick's *Do Androids Dream of Electric Sheep?* and Ridley Scott's *Blade Runner*."*Critical Survey* 3.3 (1991): 297–304.

Part IV

NEW ARRIVALS

Chapter 9

Be/Holding Each Other

Transgenic Invisibilities, Anomaly, and Subjectivity in the GFP Bunny Project

Malin Palani

I have held many animals. As a child on an Iowa farm in the 1970s and early 80s, I would lie down in the cold snow for hours with upwards of a dozen cats snuggled closely to the entire landscape of my snow-suited body. I have held living, dying, and dead animals: injured birds that could not fly, rabbits that could not hop away, foals and kittens and chicks that had just been born. I have held ducks and chickens with their heads cut off; squirrels and rabbits with their bellies gutted; frogs, fish, turtles, and deer that ended up on the family dinner table. I have held pigs with belly ruptures, blowouts, and abscesses—from Iowa hog confinements—that could not stand on their own or could not stand at all. I have carried "roadkill" home to bury and held many animals before their cremations. Much of writing this essay occurred while holding an animal.

On April 29, 2000, in Jouy-en-Josas, France, Eduardo Kac first held Alba, a rabbit who Kac genetically altered to glow fluorescent green under the right light.[1] As part of his "GFP Bunny" transgenic artwork, the image of Kac holding Alba marks a particularly potent consolidation of resonances for thinking how human relationships with nonhuman animals are made and made visible in a biotech age. While performance art has a history of challenging the boundaries and identifications of the human body, transgenic art such as "GFP Bunny" expands these investigations into genetic and more-than-human realms—in part confronting conceptions of the body not just in terms of identity but also in terms of biology and ontology. As a subgenre of BioArt, transgenic art entails the introduction of synthetic genes into an organism or the transference of "naturally" occurring genetic material from one organism to another.[2] The "GFP Bunny" project adds jellyfish DNA to the genetic makeup of a rabbit—thereby creating an animal that has two distinctively different genetic lines within one body. Eduardo Kac's transgenic alteration of a rabbit comments on human artistic and scientific involvement—however transparent or opaque—with nonhuman animals. The project draws attention to the potential for biomaterials to be reused and reconfigured in politically and ethically weighted biotechnologically enabled manipulations of living organisms, tissues, and genetic structures. Stable definitions of the body and of life are called into question along with historically rooted contexts and habits of human practices.[3] In BioArt and transgenic art, the often opaque operations of power that legislate and classify life and lives into hierarchies and taxonomies might be subjected to critique and modification.[4]

While Alba might be understood as the genetically altered organism that in part exists beyond the parameters of the project (and of a live performance), the "GFP Bunny" project includes the interactions between Eduardo Kac, the scientists, Alba, and the public; the public presentation of the animal created; the scientific procedures and socialization processes performed; the ongoing discourse around the project; Eduardo Kac's textual framing; all "GFP Bunny" media and images including that of Kac holding Alba and the emblematic green bunny; and, lest we forget, a rabbit and a jellyfish. In Kac's description, "GFP Bunny" is an ongoing social event that includes the creation of the chimerical animal, the public dialogue concerning the work, and the social integration of the rabbit.[5] Kac's "GFP Bunny" chapter in *Telepresence & Bio Art Networking Humans, Rabbits & Robots* might be considered as a frame through which the project is documented and/or as part of an elaborate public relations campaign. The discursive formations in the "complex social event" labor in part to produce the presence and visibility of a nonhuman chimera.

A chimera is a single organism that is comprised of two or more genetically distinct cellular compositions, which present a problem for DNA testing.[6] Chimeras are sometimes out of place in systems of species classification and often cannot be seen or identified without genetic testing. While biotechnologically produced organisms resultant from tissue engineering in BioArt rely on the visibility of the thing created, a chimera such as Alba undermines classifications in a nonvisible way insofar as a particular light is necessary to make visible the genetic alteration, the GFP, that makes Alba appear green. Therefore, how the chimerical animal is made visible and intelligible—whether or not under the "right" light or as a multiple in a single body—suggests the significance of visibility and visuality for the project and for a particular affect-infused relation that resonates with chimerical multiplicity.

Jacques Derrida takes up the concept of the chimera in *The Animal That Therefore I Am* and suggests that a chimera is "a sort of monstrous hybrid," whose name is derived from the multiplicity of *animot* in it.[7] Hybrid multiplicity challenges filial and linear models of evolution and underscores a body not as a relatively closed organism but as a continual process of becoming with/in the surrounding world. Gilles Deleuze and Félix Guattari theorize becoming as a multiplicity that continually transforms into other multiplicities—a plurality of heterogeneous intersections that cannot add or lose a dimension without changing nature.[8] These ongoing transformations or "strings of multiplicities" operate with the force of affect theorized for Deleuze and Guattaria's prepersonal forces that increase or decrease a body's capacity to act.[9] Affect might be understood as an entire relational field of becomings permeating and *participating* through and across all matter and all assemblages that compose moving bodies and worlds together. As a complex social event in which—genetic, social, biological, and techno-mediated—participation figures prominently in the making of the work, the "GFP Bunny" project becomes laden with implicit and explicit affect raising the questions: how are affects mobilized in "GFP Bunny" and how are the affective capacities of bodies bound to visibility and conventions of visuality in the work? Additionally, how do I be/hold multiplicity in "GFP Bunny"; how does "GFP Bunny" make visible the invisible; and how do media and knowledge formations act as mechanisms for affect as well as function to harness affect in the production of bodies, subjectivities, and disciplinary apparatuses? The well-known image of Eduardo Kac holding Alba visualizes a sense of vulnerability, which generates an unease concerning the interactions between the scientist/artist and the animal/object, between the (dominant) lover and (vulnerable) beloved. While biotech transformations might lead to new understandings of affect, new bodies and social forms, and new laws and regulatory forces—novelty cannot secure against the easy and sometimes ambiguous transition from affability to aggression.

Unlike "face-to face" interactions with BioArt that produce a "presence" through sensory affects that engender a feeling of empathy in the audience,[10] the presence of the chimerical animal that is produced in "GFP Bunny" hinges on the absence of face-to-face interaction. "GFP Bunny" participates in a discursive production that relies on the absence of who/what we generally think is present at the center of the artwork—namely Alba, a rabbit, this bunny. While the "GFP Bunny" project is not limited to the transgenic body of Alba, discourse and bodies shape, constrain, and enable each other. Michel Foucault argues that discourses are "practices that systematically form the objects of which they speak;" therefore, discourse does not identify objects as much as constitute them.[11] Insofar as the "GFP Bunny" project constructs a perspective of legitimacy and intelligibility for understanding a subject/object of knowledge, the embodied practices of seeing and holding that occur in "GFP Bunny" raise the question of what might legitimate visibility and conventions of visuality in the midst of transgenic procedures and in/visibilities that affectively and discursively produce bodies and publically mediate intimacy with those bodies (in part through intertwining practices of research, science, art, and mass media)?

For Martin Heiddegger, research is key to the essence of the modern age, which operates with a certain kind of beholding that is essential to science. In "The Age of the World Picture," Heidegger describes the visuality essential to the modern age as a kind of representing that objectifies for the purpose of gaining mastery over "what is."[12] The emergence of the world picture, which means "the world conceived and grasped as picture," signals the rise of man as *subjectum*—"the authoritative measure for all standards of measure."[13] The essence of the subjectivity of the *subjectum* belongs to the essence of man as subject. In other words, man as *subjectum* is the representer of all representing, the measurer of all measures, "[who] can determine and realize the essence of subjectivity always in keeping with the way in which he himself conceives and wills himself."[14] Heidegger contrasts the visuality of the *subjectum* with a manner of apprehending in which man must remain exposed in a modality of seeing as receiving rather than objectifying.[15] While exposure—particularly under the gaze of an animal—is at the forefront of Derrida's theorization, we are invited to look upon Eduardo Kac holding Alba and to look for Alba's genetic anomaly through the dissemination of the media image and within the frames of intersubjectivity and rabbit agency. Kac writes about Alba's agency and subjectivity within the project—as a dialogic and intersubjective social sphere in which Alba participates under a paradigm of care-infused domestication and discovery. The notion that man is the subject that determines the measure of all subjectivity complicates and resists any gesture to attribute agency and subjectivity to a nonhuman animal that participates in performance.[16]

For Eduardo Kac, the bunny's participation in the project is framed in terms of "rabbit agency." However, understanding what rabbit agency entails—as Kac directs us to do—is difficult at best and does not necessarily undo the anthropocentric social order that enables its conceptualization. Kac articulates that he constructs the project indeterminately so that:

> what human and nonhuman participants think, perceive, and do when they experience the work matters in a significant way. My answer is to make a concerted effort to remain truly open to the participant's choices and behaviors, to give up a substantial portion of control over the experience of the work.[17]

From this statement, rabbit agency might mean participation in the project not explicitly through affects but through the act of making choices and performing behaviors. Alba's choice might be framed as a behavioral action—a sort of chosen behavior—that rises out of and is always rooted in affective encounter with others and one's surroundings. However, choice as an action taken only for fulfilling one's own interests—choice as an *ability* to act in one's interests set against a reactive mode of behavior—constructs a reaction/response dichotomy that maintains conventions of visuality in which the human subject dominates the animal object.[18] In contrast, Derrida theorizes the *passive* nature of ability, argues that the question "can the animal respond?" (and its denials) are at the root of a philosophical Cartesianism,[19] and explicates René Descartes' distinction between reaction initiating from a law of nature and response from a law of freedom.[20] The Cartesian division between reaction and response marks out the realm of culture in which the human being freely acts through a mode of response-*ability*. In contrast, reaction occurs as an attribute of the nonhuman being—brute animals as machines in the natural domain—and as an automated behavior characteristic of an entire species, in which they have no choice to act in any other way. In Derrida's critique, the opposition between reaction and response is a construction that prioritizes ontological characteristics of types of beings over their potential for variation and change.

"GFP Bunny" operates in a modality of variation through the production of a uniquely in/visible plurality in a transgenic animal who holds the potential to disrupt the authoritative (seeing) positions of knowers/inventors. However, "GFP Bunny" also operates in a modality of objectification and reification in part through a determination of behavior (e.g., glowing green) as an effect of a genetic code that can, under an instrumental and bio-mediated directive, be made to vary. The entanglement of variation (webs of multiplicities) and objectification (domination) draws together the perspectives of plurality and plasticity. While plurality and plasticity both invoke variations rooted in bodily capacities, plasticity reorganizes "inflexible" variations that resist outside forces. Plasticity marks a certain kind of market/commodity driven variability (in which disposability as much as mobility bears upon biotechnologically enabled variations) that values the co-operation of flexibility (e.g., durable genetic material that can be transposed and manipulated) and inflexibility (e.g., genetic code determined and made visible in its capacity to perform). The value and meaning that biomaterials have in BioArt and transgenic art does not just have to do with financial profit and technological progress but also with how we practice care with others and in forming the self—in the midst of plurality and plasticity.[21]

In Kac's "GFP Bunny" chapter, care is articulated—in particular Kac's care for Alba—primarily under the rubric of a domestic relation. Kac writes at great length concerning the responsibility and care with which he proceeded in the project.[22] In the image of his care and love for Alba, Kac distances himself from the figure of a scientific experimenter to favor the role of *caretaker* and *inventor*.[23] Kac states:

> I have paid close attention and given careful consideration to any potential harm that might be caused. I decided to proceed with the project because it became clear that it was safe [...] green fluorescent protein is harmless to the rabbit. [...] the *GFP Bunny* project breaks no social rule: humans have played a direct role in the evolution of rabbits for at least fourteen hundred years.[24]

The persistent verification of the care undertaken on the project and in ensuring that Alba, the rabbit, was not harmed might be understood as an anticipatory gesture aimed at curbing potential backlash and protestations that might arise concerning any genetic engineering project. Jeremy Bentham's statement—"The question is not, Can they reason? nor Can they talk? but, Can they suffer?"[25]—is implicitly introduced insofar as Kac asserts that he does not want to cause any harm to the rabbit. We are encouraged to understand "GFP Bunny" and Alba through the lens of domestication as an unquestioned frame for care in human-animal relationships, which does not always or necessarily ensure care. Since we have no way of knowing what the world is for any nonhuman animal—domesticated or not—the seeming determination of an animal's needs are never completely certain, possible, nor innocent. Therefore, the motivation for a practice of care cannot rely on these kinds of determinations alone but must also emerge from the vulnerability that resonates in Derrida's exposure and in the uncertainty that the image of Kac holding Alba generates. Derrida indicates that Bentham's "can they suffer?" poses a counter-approach to thinking the question of *l'animot* through shared experiences, which might lead to compassion rather than domination characteristic in philosophical determinations and attributions of human and animal abilities. Disrupting the seemingly strict opposition between the human and the animal, Derrida poses the problem of what it means to follow and to respond in part through multiple modes of being—after, alongside, with, near, and huddled-together—that provide avenues by which "it [this cat] can look at me" and by which "I see myself seen naked under the gaze of a cat."[26] Derrida's following (*suis*)—who am I/who am I (following)—envisions a moment in which the human being is first looked at, seen rather than seeing, and operates as a kind of madness that frustrates and necessitates response (including but not limited to a reflex of shame) across borders in which care or lack thereof is embedded in visibility, affect, and structural formations of power.[27] Following for Derrida entails a philosophical practice that generates differentiated and nonoppositional articulations of more particularized relationships between humans and animals (and between nonhuman animals as well). Following might also be taken as a practice of attunement that is not about achieving distinct perception but continually opening toward the limit of sensibility and encountering with care, in Deleuze'swords, the "paradoxical existence of a 'something' that cannot be sensed ... and can only be sensed,"[28] whoever and however a something might be.

For Derrida, "can they suffer?" marks a fundamental shift from considering the power to possess—to have a manner of being or a capability—to a power of vulnerability. He writes that Bentham's question has a certain passivity, stating:

> It bears witness, manifesting already, as question, the response that testifies to a sufferance, a passion, a not-being-able. The word *can* [*pouvoir*] changes sense and sign here once one asks, 'Can they suffer?' ... [which] amounts to asking 'Can they *not be able*?' And what of this inability [*impouvoir*]? What of the vulnerability felt on the basis of this inability? What is this nonpower at the heart of power?[29]

Therefore, the philosophical framework within which animal rights awakens and stipulates our obligations and responsibilities must change with the experience of compassion in the shared vulnerability that humans have with nonhuman animals. "GFP Bunny" raises the question: how can the rights—and the life—of a rabbit be protected when the rabbit is not a citizen of the state (in the midst of shared vulnerability) and will likely never possess the capabilities of being (or visibility as) a citizen of the state? Insofar as "GFP Bunny" negotiates between the "private realm of the family" and "the public dimension of genetic engineering and public opinion," Kac calls for public respect for the life of the individual rabbit. Respect and care are necessary not just because Alba is a unique individual, who possesses a public and private life, but also because there is no essential opposition between the public and the private, between state power and non-power (vulnerability), or ability to inhabit one and not the other. The negotiation of public and private—as well as internality and externality—in "GFP Bunny" does not just concern the ambiguity and visibility of Alba's genetic modification but also the uncertainties and vulnerabilities in the formation of (my)engagement with "GFP Bunny" via perceptivity, sensibility, and knowledge.[30]

I have used many names for Alba, the rabbit, the chimerical animal, GFP Bunny as a way to point out the ease and arbitrariness with which I can name the glowing bunny and to attend to this irreplaceable being's presence and absence. Derrida confronts the problem of naming nonhuman animals by the singular *the* animal not by putting the word "animal" between quotation marks for citational analyses but by changing the singular term to *l'animot*, which Derrida deems a chimerical word "simultaneously proximal and radically foreign."[31] Derrida's *l'animot* is a plurality that draws attention to the problem of naming the animal in ways that come to appear as natural or common sense, thereby concealing their dominating and violent impacts.

The uncertainty of how suffering occurs, paired with the certainty of suffering, draws even a stone (not to mention a scarce jellyfish) into ethical considerations because suffering and violence are bound to an affective force. "Can they suffer?" attests to a painful finitude and continuity that permeates relation—an endurance in the midst of continuous exposure, a weight to be carried. Perhaps, if I were to hold Alba in my arms—if I were to hold Alba in this article, in my writing and thinking, in my body—I would become aware of a need for her comfort and health. Perhaps my humanity would reel in response to this anomalous creature. Perhaps, I would do whatever I could to provide for her continued health and happiness. Still, I could just as easily destroy her. The implication is that *how* (and which) bodies emerge from human-animal relation is caught up in the notion *that* bodies emerge in human-animal relation.

The philosophical and ethical relevance of Bentham's "can they suffer" set alongside a history of human practices of exploitation and brutality in everyday and research practices shows that it is not enough to acknowledge that an animal can suffer but that our (seeing-holding, being-seen, and being-held) practices are infused with power to create suffering and power to endure suffering with others. Naming and subjecting a nonhuman animal to the authority of one's own knowledge and subjective perceptions is bound to a discourse that frames and values the indiscriminate animal "for" the human. Insofar as the "GFP Bunny" project—including our participation—subjects the living mobility of bodies to uncriticized knowledge formation of human subjectivity and domestication, there is no rabbit. The uncertain flux of strange, anomalous, and emerging bodies is foregone. Although Kac and Alba are close domestically and even though Kac mentions that molecular biology "shows the human genome to be nothing special," Kac's proximity to Alba remains coded by a discourse of discovery and domesticity (knowledge as truth) through which control as much as collaboration weighs on intersubjective interactions.

Only in the ruptures of coherent embodiment—not just rabbit genetic modification—do the transformative potentials of the project come to the forefront. In seeing *the* human-animal relation by emphasizing the truth of discovery, breeding, and domestication—the transformative possibilities that "GFP Bunny" points towards are limited by an investment in practices that appear to stabilize the hierarchical limits between humans and nonhuman animals. However, Kac holding Alba raises an acute and tangible vulnerability and extreme uncertainty concerning who/what Alba sees and be/holds. The living relation between this rabbit and this man—a mortal and living vulnerability with which I participate—ensures that a life always depends on and is exposed to many others. Derrida suggests that the bond between a human and an animal resides in the awareness and actuality of their mortality. He writes of this suffering with compassion, "Mortality resides there, as the most radical means of thinking the finitude that we share with animals, the mortality that belongs to the very finitude of life, to the experience of compassion, to the possibility of sharing the possibility of this nonpower, the possibility of this impossibility, the anguish of this vulnerability, and the vulnerability of this anguish."[32] The anguish of vulnerability—a particularly affect laden sharing and belonging—is experienced in the mortality that humans and animals share, in the mortality of a loved one, more particularly *this* loved one,[33] and moreover in the mortality of *all* life.

Response and responsibility generate differentially and proximally from a relationally rooted compassion felt in the weight of vulnerability (being seen and exposed). The image of be/holding materialized in the "GFP Bunny" project informs Kac's statement that "one must be open to understanding the rabbit mind, and more specifically to Alba's unique spirit as an individual. [...] Understanding how the rabbit sees the world is certainly not enough to appreciate its consciousness, but it allows us to gain insights about its behavior, which leads us to adapt our own to make life more comfortable and pleasant for everyone."[34] However, the statement that rabbit consciousness cannot be understood through rabbit biology (how a rabbit sees) must go further because exposure—infused with vulnerability and compassion rather than behavioral *insight* for behavioral adaptation—is not just about (and does not necessarily lead to) making life more comfortable and pleasant for everyone. The (historical and philosophical) emphasis on and motivation for our own (human) comfort and convenience ignores the rupture between all beings (and our insights) and maintains a structure of domination. For Derrida, an "abyssal rupture" is evident insofar as there is no homogeneous continuity between what calls itself man and what man calls the animal.[35] Derrida claims that questioning whether one can free the relation of the essence of being to beings "from every living, utilitarian, perspective-making project ... such that man himself could 'let the being be'? radically concerns the *entire* question of being."[36] The role of biotechnology in manipulating bodies and biomaterials—in not letting beings be—contrasts with biotechnology that might also operate with affect, and/or with a quality of care beyond a utilitarian or tyrannical project of domination.

The "GFP Bunny" project produces a tension between biotechnology dependent on what a human knows seemingly conclusively of biomaterials and animal bodies and what cannot be known or what might be wondered in a receptive openness to and shared vulnerability with unknowable bodies/materials. To what extent am I (are we, are you) exposed and vulnerable in biotech mediations—rupturing a distinct self, a perceptibility, a discernability, a personality, what I think I know? Perhaps, "GFP Bunny" is made through human actors that manipulate nonhuman objects and genetic codes.

Perhaps "GFP Bunny" is made through a shared vulnerability realized in the compassion of an emerging encounter. Perhaps, realizing and practicing our own vulnerability might change the perspective that we have on the world in which we take up the position of the subject that legitimatizes and measures all subjectivity and subject/object positionalities.

No single animal can be comprehensively identified or determined in knowledge or ontological formations as strictly domesticated, agricultural, wild, or affect. The domestically framed nonhuman animals with whom we engage move among these categories—drawing them into question and activating a modality of exposure that ruptures our seemingly determined relationships. Deleuze and Guattari outline three treatments of animals—as pets, scientific classifications, and affect animals—that are not mutually exclusive.[37] In order to examine the question of the bestiality of man (bestiality as cruelty) and its relationship to sovereignty, Derrida takes up the Deleuzian critique of psychoanalysis in which "these *bêtises* of psychoanalysis are not only poverties of knowledge, nonknowledges, or incomprehensions, but ethical violences, machines in turn, and machines of war, subjection, brutalization, ways of making patients more *bêtes*, more brutal or brutish than they are in truth."[38] For Derrida, the logic of a sovereign responsibility (e.g., humanity capable of bestiality) set against reactional irresponsibility (e.g., animality incapable of bestiality) upholds in a "language belonging to accusation" the opposition between human and animal. However, *bêtise* is engendered "in the very gaze" that gives it attention with a kind of contagion that ensures *bêtise* is "never on one's own," whose plurality raise doubts about authority.[39]

General and indeterminate (seemingly determinate) categorical treatments of humans and animals generate irreducible proliferations of meanings that cannot be contained. Animals can be treated through multiple lenses of perceptivity that open to "first being seen." Perhaps I could do right by Alba without allowing this rabbit to resonate within or rupture me—constricting the bunny to a truth of normativized domestication. For my exposure to materialize, my being-becoming would have to break open for another to inhabit my body—like a jellyfish inside a rabbit—forcing my words and understanding to come unhinged. Risking potential harm—like a rabbit inside a jellyfish. Moving through inside and outside. Not to destabilize, redeploy, or recuperate subjectivity in terms of a still human-centric intersubjectivity, but to be brought to the very threshold of subjectivity itself. Here, exposure vitalizes a be/holding *with* others that points out the tension between the hierarchical practical and ontological competitions we construct and the unfolding deeply entangled affecting and affected embodiments. This shared, carved up, *with* an animal is a becoming that does not erase differences and is never reducible to stable social roles and genetic modifications—an alchemical and affective rather than inventive operation.

In "GFP Bunny," Eduardo Kac's intentions through an inventive and open approach are "to accept the experience as-it-happens as a transformative field of possibilities, to learn from it, to grow with it, to be transformed along the way."[40] But how are affects and variations mobilized in Eduardo Kac? What are those transformative potentials? What are the implications of Kac injecting himself with rabbit DNA? Might there be in any way a jellyfish-rabbit inside Eduardo Kac? Kac's vulnerability begins to show in the image of holding that permeates the project. The transformation of course does not occur in terms of a scientifically enabled genetic modification made within a discourse of the discovery or domestication of the individual named Eduardo Kac. The transformational vulnerability of Kac challenges the notion that the individual is formed only through varying genetic combinations, subjectivizations, and objectifications. The conception of the (human or nonhuman animal) individual as a formation that can be known and determined through DNA structures eliminates a vast array of processes that impact the formation of the individual including reproduction, the ongoing contingencies of relation, and the complexities and mobility of difference that cannot be simply located in a genetic or informational code. Understanding the transformation of the individual primarily in terms of genetic differences limits how we come to understand the complexities of relation—as if the individual is only and primarily a scientifically and technologically produced entity.

While Alba upsets rabbit filiations and classifications, an emphasis on the transgenic *characteristic* in Alba reduces human-animal relation to a science of genetic manipulation and genetic structure. Kac be/holding Alba (and who/what Alba be/holds) materializes a sense of transformation that does not indicate or display any genetic modification but resonates with a capacity for and exposure to affect. Affect moves through modalities of exposure in which an individual is caught up in processes of transformation and relational flux—as if a rabbit could say, "I am a jellyfish" and a prairie could say, "I am the sky," and Kac could say, "I am a rabbit," or the dust could say "I am a writer"—each becoming less or more another. Affect makes the "more" of living materials in biotech, personal, and performance practice vaguely and yet substantially tangible. What a body is capable of doing is not an inalterable fact, subjective perception, or codified genetic structure but rather the movement of affect through pathways of vulnerability. The transference of jellyfish DNA in the project is perhaps the most visible (mediated, scientific, and affective) transformation and exposure. However, the visibility of the biotech procedure is infused with the uncertainty of how green fluorescent protein is obtained or manufactured and with the uncertainty surrounding the implications and affects of genetic transference.

The project begs the question: how does a jellyfish emerge from and participate in the project? It should not go unnoticed that a jellyfish is far more difficult to convey and display compassion for—a jellyfish is far more difficult to enter into a domestic relation with. A jellyfish would not have made a good social subject or subject to care for within the domestic domain. A jellyfish is difficult to hold in a photograph. The GFP (green fluorescent protein) gestures towards a real jellyfish participating in and emerging from the project—not just a bit of engineered DNA or manufactured protein as a representative characteristic of all jellyfish. Herein, Alba and this jellyfish are beings that see us and refuse to be conceptualized—or reduced to genetic code—in the same way that *this* cat seeing Derrida cannot be conceptualized or reduced to a genetically determined and identified individual. Rather than feed a visuality that seeks to master the known and the unknown—the visible and the invisible—the refusal of any being/body/materials to be conceptualized has the potential to revitalize an artistic-scientific practice that critiques social, technological, and subject-object organizations of reality.

Biotechnology has opened up a historically specific way of understanding bodies as processes—environmentally adapting and adaptable units of information—that can be taken up in projects of intervention in matter and the domain of affect. The biomedial relation between matter and information—which Eugene Thacker understands as a technical reconditioning of biology (of which DNA is only one part) enabling biology to perform in novel ways[41]—introduces tension between information as affect laden connectability and information as decipherable code. The biotech alteration of a species—and a particular rabbit body—makes visible and manipulable materials that perhaps require us again to ask "can they suffer?" In the midst of biomaterials that compose symbiotic assemblages, what does it matter and mean for us to ask "can they suffer?" The prepersonal flow of affect might be engaged as a shared field of exposure and/or as a field to be instrumentally engaged for the extraction of value. How might biomaterials *see first* the human subject (property owner, responsible consumer, inventive entrepreneur, care(less) provider, etc.), who is encouraged to pursue value with an unquestioning faith in technology for the production of a "global good" that cannot be disentwined from a project of (market driven) domination?

In describing the exposure felt at being stared at naked by an unnamed cat, Derrida explicates that *this* cat, not *my* cat, does not represent the entire species cat but appears as a living being with "unsubstitutable singularity."[42] This cat, this rabbit, this jellyfish mark mobile limits with which and by which one might be stripped naked, profoundly interrogated, and effaced. Derrida suggests that the gaze and the classification called "animal" offer the abyssal limit of the human, which he describes as the "border crossing from which vantage man dares to announce himself to himself."[43] In moments of exposure in transgenic art, naked under the gaze of a pluralized animal, as well as under the eyes of the strange reader/s of my writing, anything can happen, but what of the gaze of GFP? The "apocalyptic" encounter with *l'animot* that Derrida describes conveys an ethics of exposure rather than a set of rules for the balancing of competitions and interests and an ordering for mutual benefit. Derrida gestures toward this ethics where he elsewhere outlines "the ordeal of the undecidable" that must be endured—a sufferance of exposure and a permanent and relentless condition of rupture.[44] Derrida's theorization of undecidability articulates the disruption of binary structures that attempt to define the limits of living relations. Moments in which one finds or follows oneself naked under the gaze of—in the flow of affect with—an animal, the opposition between the dominant (lover) subject and the vulnerable (loved) object is displaced toward an undecidability rooted in a visuality in which the dominant is seen exposed and therefore necessarily affected in the most subtle, the most striking, and the most unspeakable ways.

In the "GFP Bunny" project, exposures of the human are enlivened in the vigilance of listening (attending and adjusting) to something that is not immediately accessible and that unfolds with vague perceptions .Listening as a mode of visuality is ultimately not about comprehensive and stable perception of an animal—nor instrumental calculation in the midst of uncertainty—but about vulnerable, embodied, and sensorial stretching with concern and responsiveness rather than response-ability. Jean-Luc Nancy argues that, "To listen is *tender l'oreille*—literally, to stretch the ear … it is an intensification and concern, a curiosity or an anxiety."[45] To stretch the ear intensely—*to stretch the very skin and viscera of the entire body*—is to become awake with curiosity or anxiety and concern and "to be straining toward a possible meaning, and consequently one that is not immediately accessible."[46] Waiting, opening, paying attention … *keep* waiting … listening while naked materializes a temporal spatial embodied vulnerability.

Perhaps I have not strained enough to listen for the affect-emerging transformations occurring in the "GFP Bunny" project. Perhaps I did not shamefully see myself being seen naked. Sometimes a human writes, speaks, and publishes so as not to listen. Some nonhuman animals, perhaps such as a rabbit, seem to be always listening—becoming all ears so-to-speak, always on the aural lookout, always (at)tuning themselves to their surroundings—always seeing me. While Alba's aberrance appears in part as a genetic code or a characteristic that makes this rabbit an anomaly, the individuality of all animals must be listened to, for, within the midst of mobility and naked uncertainty. Deleuze and Guattarri argue:

> It is clear that the anomalous is not simply an exceptional individual; that would be to equate it with the family animal or pet … . Nor is the anomalous the bearer of a species presenting specific or generic characteristics in their purest state; nor is it a model or unique specimen … . The anomalous is neither an individual nor a species; it has only affects, it has neither familiar or subjectified feelings, nor specific or significant characteristics.[47]

Whether the "GFP Bunny" project exposes the human, demonstrates genetic transformations that have the potential to affect, and/or attempts to capture the living heart of an animal in performance (perform and else!)[48]—a practice of care in which we might be/hold others and are first be/held by others (including biomaterials in symbiotic assemblages) is not limited to organisms that breathe or see scientifically. In order to cultivate openness and vulnerability in exposure, material practices—including my writing with "GFP Bunny"—require patience, attention, generosity, respect, and humility in the midst of the many piercings (visual and otherwise) that have the potential to transform.

NOTES

1. The photo image by Chrystelle Fontaine that I would like to introduce the project with can be viewed online at http://www.ekac.org/gfpbunny.

2. While Eduardo Kac coined the term "BioArt" in 1997, the biotechnologically materialized modifications in BioArt are made possible through a range of scientific processes such as cloning, hybridization, genotype and phenotype reprogramming, and genetic and tissue culture engineering. BioArt not only includes the manipulation of animal genetic material through technologies found in the biological sciences but also the manipulation of plant material through technologies in the plant sciences.

3. Deborah Dixon, "Creating the Semi-living: On Politics, Aesthetics and the More-than-human," *Transactions of the Institute of British Geographers* 34, no. 4 (Oxford: Blackwell, 2009), 411–25.While some BioArt projects convey a particular and more explicit political message tied to specific techno-scientific practices, other works take a more ambiguous approach in their political activity. Deborah Dixon argues that BioArt inherently performs a political function by reconfiguring social order and reordering subjects and objects.

4. Oron Catts and Ionat Zurr, "The Ethics of Experiential Engagement with the Manipulation of Life," *Tactical Biopolitics: Art, Activism, and Technoscience*, ed. Beatriz da Costa and Kavita Philip (Boston, MA: MIT Press, 2008), 140.BioArt raises questions concerning medical and pharmacological experimentation and research, re/productive and genetic modification of plants and animals, and rights of bio-ownership. BioArt is embedded in the practices that it seeks to critique. When DNA becomes determined as and reduced to an informational code and when living tissues become "mere" tissue—which can be endlessly and easily manipulated—the new ungraspable understandings of life that BioArt gestures towards can be reduced to one understanding that is a delimitation of life as informational code. Even if intentioned as sociopolitical critique and awareness—the reduction, reading, and breaking of living materials as code for science or art does not necessarily operate against power structures but can reinforce and normativize power structures in a form of techno-scientific domination that operates through the manipulation of life considered as genetic resource. The question of how BioArt is *ethically* motivated is perhaps the most controversial.

5. Eduardo Kac, "GFP Bunny," *Telepresence & Bio Art Networking Humans, Rabbits, & Robots* (Ann Arbor, MI: The University of Michigan, 2005).

6. Neng Yu, *et al.* "Disputed Maternity Leading to Identification of Tetragemetichimerism," *New England Journal of Medicine* 346, 20 (16 May 2002): 1545–52.Sylvia Pagán Westphal, "'Humanised' Organs can be Grown in Animals," *New Scientists* (17 December 2003). Human chimeras can occur as in the case of Lydia Fairchild who was found in 2002 to have two sets of DNA after being presented with a set of DNA evidence that showed that she was not related to her children. Chimeras also occur through procedures such as organ transplant and more controversial procedures such as the selective transplantation of embryonic cells from one organism onto the embryo of another. In 2007, the University of Nevada School of Medicine produced a chimera with human and nonhuman DNA—a sheep with 15% human cells and 85% sheep cells.

7. Jacques Derrida, *The Animal That Therefore I Am*, ed. Marie-Louise Mallet, trans. David Wills (New York, NY: Fordham University Press, 2008), 42.

8. Deleuze and Guattari's theorization of becoming draws in part on Friedrich Nietzsche's philosophical formulation of becoming and on David Hume's theorization of the nature of relation in which relation occurs *exterior* to the distinct terms that interact. Relation occurs between and heterogeneous to distinct terms—operating with a principle of movement rather than correspondence. Therefore, becoming is not a correspondence between distinct terms, subjectivities, or identifications and relation is not the result of distinct terms interacting but precedes and exceeds their interactions and correspondence. Deleuze and Guattari write, "Since its variations and dimensions are immanent to it, *it amounts to the same thing to say that each multiplicity is already composed of heterogeneous terms in symbiosis, and that a multiplicity is continually transforming itself into a string of other multiplicities, according to its thresholds and doors.*" In Gilles Deleuze and Félix Guattari, *A Thousand Plateaus Capitalism and Schizophrenia*, trans. Brian Massumi (Minneapolis, MN: University of Minnesota Press, 1987), 249.Also see Brian Massumi, *Parables for the Virtual Movement Affect Sensation* (Durham, NC: Duke University Press, 2002).

9. Deleuze and Guattari explain that affect is, "an ability to affect and be affected. It is a prepersonal intensity corresponding to the passage from one experiential state of the body to another and implying an augmentation or diminution in that body's capacity to act ... (with body taken in its broadest possible sense to include 'mental' or ideal bodies)." Affect involves a "degree" entering into composition with "another degree" to form another distinct individuality, which is irreducible to a subject since individuality always forms another individuality in another degree of participation. Deleuze and Guattari, *A Thousand* xvi, 253.

10. Jens Hauser, "Observations on an Art of Growing Interest: Towards a Phenomenological Approach to Art Involving Biotechnology," *Tactical Biopolitics: Art, Activism, and Technoscience*, ed. Beatriz da Costa and Kavita Philip (Boston, MA: MIT Press, 2008), 87.

11. Michel Foucault, *Archaeology of Knowledge*, trans. A. M. Sheridan Smith (London: Tavistock, 1972), 49–54.

12. Martin Heidegger, *The Question Concerning Technology and Other Essays*, trans. William Lovitt (New York, NY: Harper and Row, 1977), 127, 134, 148–52.

Heidegger refers to this representing as "enframing," in which man as the arbiter and knower of knowledge decides "by himself and for himself, what, for him, should be 'knowable' and what knowing and the making secure of the known, i.e., certainty, should mean." The essence of technology as a way of revealing in the modern age is tied to enframing [*Ge-stell*]—a self-revealing—that challenges-forth the world as a picture. Heidegger states, "The fundamental event of the modern age is the conquest of the world as picture. The word 'picture' [*Bild*] now means the structured image [*Gebild*] that is the creature of mans' producing which represents and sets before. In such producing, man contends for the position in which he can be that particular being who gives the measure and draws up the guidelines for everything that is." Herein, technology functions as a representationalist way of revealing in which the world is constructed as an object of representational knowledge where man is the only and entire subject. Heidegger argues that through modern technology man's challenging [*Herausfordern*] "puts to nature the unreasonable demand that it supply energy which can be extracted and stored as such" for maximum yield at minimum expense. In Martin Heidegger, "The Question Concerning Technology," *Basic Writings from Being and Time (1927) to The Task of Thinking (1964)*, ed. David Farrell Krell (New York, NY: Harper & Row, Publishers, 1977), 296. What Heidegger describes as modern technology's inauthentic mode of revealing challenges forth the world in a way that conceals the way of revealing of truth. This way of revealing—in contrast to a representationalist way—is not the correspondence of a representation to an objective or subjective reality but *poiesis*—a bringing-forth of the world in a revealing *that also conceals*. In critique of Heidegger's philosophical use of terms such as "Dasein" and "care," Theodor Adorno argues that after World War II the use of existential terms evident in Heidegger's work generates a jargon of authenticity that obscures the relation between language and objective content. Adorno emphasizes that idealized symbols of the jargon of authenticity do not represent actual social relations but rather abstract relations between concepts that appear to remain untouched by processes of abstraction and history. Therefore, Heidegger's phenomenology and ontological language feeds on the opposition between the authentic and the inauthentic, which reinforces knowledge of the subject as an object of explicit calculation and pre-established knowledge formations. In Theodor Adorno, *The Jargon of Authenticity*, trans. Knut Tarnowski and Frederic Will (London: Routledge, 1973).

13. Heidegger, *The Question*, 129, 139.Heidegger argues that research and the publishing business are foundational to the modern "world picture" and that they work together "to bring the world into the picture for the public and to confirm it publicly."

14. Ibid., 152.

15. Ibid., 131;Heidegger contrasts modern representing as an objectifying visuality based in subjective perception—in which the "world as picture" comes into being "through the fact that man first looks upon it"—with Greek apprehending in which man is looked upon by that which is and "is borne along by it."

16. Michel Foucault, *The Hermeneutics of the Subject Lectures at the Collège de France 1981–1982*, ed. Frédéric Gros, trans. Graham Burchell (New York, NY: Picador, 2005), 17. Michel Foucault interprets modernity as characterized by a process in

which discourses of knowledge increasingly police subjects. Foucault states, "I think the modern age of the history of truth begins when knowledge itself and knowledge alone gives access to the truth. That is to say, it is when the philosopher (or the scientist, or simply someone who seeks the truth) can recognize the truth and have access to it in himself and solely through his activity of knowing, without anything else being demanded of him and without him having to change or alter his being as subject." Modernity thereby indicates a sociohistorical situation in which knowledge takes over as the source of truth. Power then is not just a matter of domination or oppression but is productive of discourses of truth and power over bodies that can be managed more effectively and efficiently when a subject is constituted and seen as a subject of knowledge as truth.Herein, knowledge formations aim to constitute and define bodies through an exercise of power over them. Power then is not just a matter of domination or oppression but is productive of discourses of truth, which can be critiqued.

17. Kac, *Telepresence*, 272.

18. The dichotomy between reactive behaviors and responsive choices—rooted in the freedom and rights of the human subject—permeates not only philosophical distinctions between humans and animals but also understandings of how a human and an animal might participate in performance as evident in the notion that human creators (actors, directors, designers, writers, etc.) shape performance while more-than-human matters and bodies are subservient to what human creators think and do.

19. Derrida, *The Animal*. Tracing a hegemonic Cartesianism through philosophical discourses, Derrida explains how Kant, Levinas, Heidegger, and Lacan each in his own way deny the animal's capacity to respond. Kant reaffirms the human as the rational animal based upon an "I," which is a presence to oneself. Levinas leaves the animal outside of ethical consideration by denying the animal a face and making the responsive face originally human. Lacan opposes human response to animal reaction reinforcing the animal's inability to deceive by way of an original perfection. Finally, Heidegger theorizes the animal as being poor-in-the-world and the human as world-forming. Derrida shows that in Heidegger's comparative strategy, the animal does not have the world "as such." The animal does not exist or die. It has no Daesin. It merely lives. Derrida demonstrates these theories and their contradictions and failings and asserts that it is essential to get beyond the opposition between the world and the world "as such." For Derrida these understandings of "the animal" hinge on how response is defined in Western philosophical traditions. Grounding the question concerning the agency of the subject in the ability to respond or react through a particular vein in Western philosophy enables Derrida to critique philosophical framing of "the animal" without more broadly addressing the problem of the subject throughout philosophical discourse. One trajectory that Derrida does not take up in his study, which emphasizes variation and change, is Friedrich Nietzsche's theorization of the formation of subjectivity through a web of power that continually fluctuates according to competing interests and the resistances that it generates. In an investigation of the human, Nietzsche proposes that there can be no universal account of humanity or of human history since there is no human essence that can be located in principles such as subjectivity, liberty, and equality that align and consolidate across all human societies. Nietzsche argues that judgments about man derive from limited periods

in time that cannot account for the fact that everything becomes rather than just is. Therefore, there is a need for a "modest historical philosophizing." Nietzsche writes, "there are no eternal facts, just as there are no absolute truths." In Friedrich Nietzsche, *Human, All-Too-Human*, trans. R. J. Hollingdale (Cambridge: Cambridge University Press, 1986), section 2. Nietzsche also argues that humanity should be understood as "an inextricable multiplicity of ascending and descending life-processes … [in which] the strata are twisted and entwined together." Therefore, the constitution of subjects occurs in a web of power wherein different societies have different manifestations of normative subjects and resistances. In Nietzsche's theorization, society imposes prescriptions that constitute and regulate the normative subject—which is an expression of the power relations in society. In this social paradigm, power is not the possession of an individual agent but an ongoing negotiation in a social order where power continually shifts among continually forming subjectivities. Relations of force rather than abilities constitute domains of subjectivity. Therefore, subjectivity is not the ground of agency or the subject but is formed and reformed through fluctuations in power. In Friedrich Nietzsche, *The Will to Power*, trans. Walter Kaufmann and R. J. Hollingdale (New York, NY: Vintage Books, 1968), section 339.

20. René Descartes, "Discourse on the Method," *The Philosophical Writings of Descartes*, vol. 1, trans. John Cottingham, Robert Stoothoff, and Dugald Murdoch (Cambridge, MA: Cambridge University Press, 1985), 139–40. In Descartes' philosophy concerning human and animal capabilities, animals are capable of emitting signs but are not capable of responding. Descartes argues that animals like machines cannot use words like a human being—who produces different arrangements of words according to his thoughts in order to convey a particular meaning to another human being.

21. Lori Andrews and Dorothy Nelkin, *Body Bazaar: The Market for Human Tissue in the Biotechnology Age* (New York, NY: Crown, 2001), 7; Andrews and Nelkin argue that, "Definitions of the body that reduce and decontextualize it, are what allow scientists or biotechnology firms to extract, use, and patent body tissue without reference to the individual or consideration of his or her personal desires and social needs." In the midst of our nearly incessant reworkings of animal attributes and capabilities—whether considered as subject or object—many nonhuman animals embody and experience marginalization and exploitation. Likewise, cells and other biomaterials extracted from a body do not become mere subjects or mere objects emptied of history and memory made ready for use. The tie to "where things come from" is not just an issue of rights and ownership but also of maintaining the resonances that are carried in all living materials. Removed and detached from histories, lives, and experiences beyond the walls of the laboratories and galleries—we do not fully understand the in/visible transformations and exchanges of biomaterials in our biotechnological age.

22. Kac, *Telepresence*, 264.Kac stipulates, "Transgenic art … is a new art form based on the use of genetic engineering to create unique living beings. This must be done with great care; with acknowledgment of the complex issues thus raised; and, above all, with a commitment to respect, nurture, and love the life thus created."

23. Recent job postings for "Caretaker I" and "Caretaker II" at a local university listed among the job responsibilities: "restraint of animals for the purposes of experimentation" and "euthanize animals as directed."

24. Kac, *Telepresence*, 273.

25. Jeremy Bentham, *An Introduction to the Principles of Morals and Legislation* (Mineola, NY: Dover Publications, 2007).

26. Derrida, *The Animal*, 10–11.

27. The phrase and title "Seeing Animals" can be read in at least two ways: we "humans" as the unwritten subject of the phase engaged in seeing animals through the work collected here and/or "Seeing" as an adjective that describes the doing of animals, who somehow (differentially and non-oppositionally) permeate and see us through, in, and with our embodied writings.

28. Gilles Deleuze, *Difference and Repetition*, trans. Paul Patton (New York, NY: Columbia University Press, 1968/1994), 236.

29. Derrida, *The Animal*, 27–28.

30. Jean-Luc Nancy, *The Inoperative Community*, ed. Peter Connor, trans. Peter Conner, Lisa Garbus, Michael Holland, and Simona Sawnhey (Minneapolis, MN: University of Minnesota Press, 1991), xxxvii. Jean-Luc Nancy claims, "'To be exposed' means to be 'posed' in exteriority, according to an exteriority, having to do with an outside *in the very intimacy* of an inside." Further, my writing is exposed to this rabbit—a shared and differentiated exposure that potentially transforms.

31. Derrida, *The Animal*, 40–41.

32. Ibid., 28.

33. Love/grief and death between a human and a nonhuman animal often resides on the margins—met with silence and a lack of understanding—in part because understanding this love and loss is often thrust against discerning human eyes and normatizing forces that determine what is acceptable or not, what can be said of the relation or not, and what can be felt in the relation or not often qualified in terms of human relationships with other humans. Working at Memories Pet Loss Center in Iowa, I engaged with families who had lost a pet loved one—and who exhibited a measure of shame over their grief and an apologetic approach to expressing the love that they shared with their nonhuman companions.

34. Kac, *Telepresence*, 273–74.

35. Derrida, *The Animal*, 30–51.

36. Ibid., 160.

37. Deleuze and Guattari, *A Thousand*, 240–41, 244.

38. Jacques Derrida, *The Beast and the Sovereign Volume I*, ed. Michel Lisse, Marie-Louise Mallet, and Ginette Michaud, trans. Geoffrey Bennington (Chicago, IL: The University of Chicago Press, 2009), 140–44.

39. Ibid., 158.

40. Kac, *Telepresence*, 272.

41. Eugene Thacker, *Biomedia* (Minneapolis, MN: University of Minnesota Press, 2004), 12–15.

42. Derrida, *The Animal*, 9.

43. Ibid., 12.

44. Jacques Derrida, "Force of Law: The 'Mystical Foundation of Authority,'" *Deconstruction and the Possibility of Justice*, ed. Drucilla Cornell, Michel Rosenfeld, and David Gray Carlson, trans. Mary Quaintance (London: Routledge, 1992), 24. For

Derrida, "A decision that did not go through the ordeal of the undecidable would not be a free decision, it would only be the programmable application or unfolding of a calculable process. It might be legal; it would not be just."

45. Jean-Luc Nancy, *Listening*, trans. Charlotte Mandell (New York, NY: Fordham University Press, 2007/2002), 5, 21–22, 31, 38. Nancy describes body vibrating with listening as: "the rhythmic reployment/deployment of an enveloping between 'inside' and 'outside,' or else folding the 'outside' into the inside,' invaginating, forming a hollow, an echo chamber or column, a resonance chamber." For Nancy, the subject might be seen as the part of the body that is listening or vibrates with listening to or with the "echo of beyond-meaning." Even though Nancy theorizes subjectivity as the part of the body that vibrates with listening, he also writes that this subject is perhaps not a subject "except as the place of resonance."

46. Ibid., 6, 32;Nancy explicates, "If *listening* is distinguished from *hearing* both as its opening (its attack) and as its intensified extremity, that is, reopening beyond comprehension (of sense) and beyond agreement or harmony (*harmony* [entente] or *resolution* in the musical sense), that necessarily signifies that listening is listening to something other than sense in its signifying sense."

47. Deleuze and Guattari, *A Thousand*, 240–41, 244.

48. Jon McKenzie, *Perform or Else: From Discipline to Performance* (London: Routledge, 2001). Jon McKenzie theorizes an "onto-historical formation" that is bound together in technological, organizational, and cultural paradigms of performance and which together create a threatening mandate for subjects to "perform *or* else" be fired, outdated, downsized, marginalized, etc. I suggest that "perform or else" becomes "perform and else" under a formation and paradigm of plasticity in which subjects, bodies, and materials perform *and* are continually under the force of durable and disposable variation.

WORKS CITED

Adorno, Theodor. *The Jargon of Authenticity*. Translated by Knut Tarnowski and Frederic Will. London: Routledge, 1973.

Andrews, Lori Andrews and Dorothy Nelkin. *Body Bazaar: The Market for Human Tissue in the Biotechnology Age*. New York, NY: Crown, 2001.

Bentham, Jeremy. *An Introduction to the Principles of Morals and Legislation*. Mineola, NY: Dover Publications, 2007.

Catts, Oron and Ionat Zurr. "The Ethics of Experiential Engagement with the Manipulation of Life," in *Tactical Biopolitics: Art, Activism, and Technoscience*, 125–142. Edited by Beatriz da Costa and Kavin Philip. Boston, MA: MIT Press, 2008.

Deleuze, Gilles. *Difference and Repetition*. Translated by Paul Patton. New York, NY: Columbia University Press, 1968/1994.

Deleuze, Gilles and Félix Guattari. *A Thousand Plateaus Capitalism and Schizophrenia*. Translated by Brian Massumi. Minneapolis, MN: University of Minnesota Press, 1987.

Derrida, Jacques. "Force of Law: The 'Mystical Foundation of Authority,'" In *Deconstruction and the Possibility of Justice*, 3–67. Edited by Drucilla Cornell, Michel Rosenfeld, and David Gray Carlson. Translated by Mary Quaintance. London: Routledge, 1992.

———. *The Animal That Therefore I Am*. Edited by Marie-Louise Mallet. Translated by David Wills. New York, NY: Fordham University Press, 2008.

———. *The Beast and the Sovereign Volume I*, Edited by Michel Lisse, Marie-Louise Mallet, and Ginette Michaud. Translated by Geoffrey Bennington. Chicago, IL: The University of Chicago Press, 2009.

Dixon, Deborah. "Creating the Semi-living: On Politics, Aesthetics and the More-than-human," *Transactions of the Institute of British Geographers* 34, no. 4 (Oxford: Blackwell, 2009): 411–25.

Fehilly, Carole, S. M. Willadsen, and Elisabeth Tucker."Interspecific Chimaerism between Sheep and Goat," *Nature* 307 (16 February 1984): 634–636.

Foucault, Michel. *Archaeology of Knowledge*. Translated by A. M. Sheridan Smith. London: Tavistock, 1972.

———. *The Hermeneutics of the Subject Lectures at the Collège de France 1981–1982*. Edited by Frédéric Gros, Translated by Graham Burchell. New York, NY: Picador, 2005.

Hauser, Jens. "Observations on an Art of Growing Interest: Towards a Phenomenological Approach to Art Involving Biotechnology," In *Tactical Biopolitics: Art, Activism, and Technoscience*, 83–104. Edited by Beatriz da Costa and Kavita Philip. Boston, MA: MIT Press, 2008.

Heidegger, Martin. *The Question Concerning Technology and Other Essays*. Translated by William Lovitt. New York, NY: Harper and Row, 1977.

———. "The Question Concerning Technology," *Basic Writings from Being and Time (1927) to The Task of Thinking (1964)*. Edited by David Farrell Krell. New York, NY: Harper & Row, 1977.

Irigaray, Luce. *An Ethics of Sexual Difference*. Translated by Carolyn Burke and Gillian C. Gill. Ithaca, NY: Cornell University Press, 1984.

Kac, Eduardo. "GFP Bunny," In *Telepresence & Bio Art Networking Humans, Rabbits, & Robots*. Ann Arbor, MI: The University of Michigan, 2005.

Lefebvre, Henri. *The Production of Space*. Oxford: Blackwell, 1991.

Massumi, Brian. *Parables for the Virtual Movement Affect Sensation*. Durham, NC: Duke University Press, 2002.

McKenzie, Jon. *Perform or Else: From Discipline to Performance*. London: Routledge, 2001. Nancy, Jean-Luc. *The Inoperative Community*. Edited by Peter Connor. Translated by Peter.

Naas, Michael, *The End of the World and Other Teachable Moments Jacques Derrida's Final* by John D. Caputo. New York, NY: Fordham University Press, 2015.

Nancy, Jean-Luc. The Inoperative Community. Edited by Peter Connor. Translated by Peter Connor, Lisa Garbus, Michael Holland, and Simona Sawnhey. Minneapolis, MN: University of Minneapolis Press, 1991.

———. *Listening*. Translated by Charlotte Mandell. New York, NY: Fordham University Press, 2007/2002.

Nietzsche, Friedrich. *The Will to Power*. Translated by Walter Kaufmann and R. J. Hollingdale. New York, NY: Vintage Books, 1968.

———. *Human, All-Too-Human*. Translated by R. J. Hollingdale. Cambridge: Cambridge University Press, 1986.

Neng Yu, *et al.* "Disputed Maternity Leading to Identification of Tetragemetichimerism," *New England Journal of Medicine* 346, no. 20 (16 May 2002): 1545–52.

Stewart, Kathleen. *Ordinary Affects*. Durham, NC: Duke University Press, 2007.

Thacker, Eugene. *Biomedia*. Minneapolis, MN: University of Minnesota Press, 2004.

Westphal, Sylvia Pagán. "'Humanised' Organs can be Grown in Animals," *New Scientists* (17 December 2003).

Chapter 10

The Surreal Gaze of the Animal Other

Uncanny Encounters in Magritte and Buñuel

Kirsten Strom

The uncanny is typically described as an experience of that which is simultaneously strange and familiar. The cognitive dissonance that arises from this contradiction may generate anxiety and even fear or dread. The gaze of Derrida's cat, as another within the home, seemed to provoke in Derrida an experience of the uncanny, though Derrida did not himself invoke the term in this context. His elaborate self-reflection on the experience, however, exhibits a willingness to confront that which is disarmingly uncanny rather than to disavow it. In this, he was perhaps outdone by the Surrealists before him, who not only confronted the uncanny but actively courted it in their work. Though the uncanny in Surrealism took a great many forms, there are significant instances in which nonhuman animals seem to trigger the experience in ways that anticipate aspects of Derrida's later arguments.

Freud's essay on "The Uncanny" (*Das Unheimlich*) of 1919 was an important touchstone for the Surrealists, who treated it almost as if it were a manual. In Freud's discussion, nonhuman animals do not play a significant role, though there are two references to them. The first is contained within Freud's quotation of Daniel Sanders's definition of the *Heimlich*:

a. (Obsolete) belonging to the house or the family
b. Of animals: tame, companionable to man. As opposed to wild, for example, "Animals which are neither wild nor *heimlich*," etc. "Wild animals ... that are trained to be *Heimlich* and accustomed to men." "If these young creatures are brought up from early days among men they become quite *Heimlich*, friendly, etc."—So also: "It (the lamb) is so *Heimlich* and eats out of my hand." "Nevertheless, the stork is a beautiful, *Heimlich* bird."
c. Intimate, friendlily comfortable.[1]

Following the translation of *unheimlich* into *uncanny*, the English version of *heimlich* is *canny*. The German terms, however, evoke a more direct association with of the idea of home; the *heimlich* is that which is homelike, and the *unheimlich* that which is un-homelike. This is evident in Sanders's definitions, which link the house, the family, the intimate, and the animal that has been tamed. What is telling, however, is that even that animal which is not properly domestic, in the sense of being of the home, such as the stork, may be called *heimlich* so long as it is nonthreatening. This suggests that the comfort Sanders described stems from a sense of an animal that knows its place, even if that place is not strictly speaking in the home. Freud did not develop the implication that "wild" animals might be experienced as *unheimlich*, but there is also here potential that even the most domesticated of animals might also be uncanny if it upsets the comfort of a familiar, accepted, and implicitly hierarchical relationship between human and nonhuman animals.

The second (and final) reference to other animals in the essay appears in Freud's description of

> a story about a young married couple who move into a furnished house in which there is a curiously shaped table with carvings of crocodiles on it. Towards evening ... they stumble over something in the dark; they seem to see a vague form gliding over the stairs—in short, we are given to understand that the presence of the table causes ghostly crocodiles to haunt the place, or that the wooden monsters come to life in the dark. ... It was a naïve enough story, but the uncanny feeling it produced was quite remarkable.[2]

What is uncanny for Freud in this instance, however, is not the crocodile per se, but rather its ghost, or alternately, the sinister magic that collapses the distinction between an inanimate representation and the animate life form that it represents. Indeed, inanimate representations of the animate, namely dolls, mannequins, and automatons play an important role in Freud's discussion of the uncanny. Their importance is detailed in a lengthy reference to Ernst Jentch's 1906 reading of E. T. A. Hoffman's story "The Sand-Man." Jentch had concluded that the story's uncanny effects were generated by the ambiguous nature of the automaton Olympia that becomes the love object of the story's protagonist, who mistakes the object for a real woman. According to Freud, "Jentsch has taken as a very good instance 'doubts whether an apparently animate being is really alive; or conversely, whether a lifeless object might not be in fact animate'; and he refers in this connection to the impression made by wax-work figures, ingeniously constructed dolls and automata."[3] Derrida similarly remarked upon the ambiguousness of the marionette invoking this very term: "The marionette—*who* or *what*? And what if it ... were between the two ... between the who and the what—both sensible and insensible, neither sensible nor insensible ... living dead, spectral, uncanny, *unheimlich*?"[4]

Dolls and mannequins feature prominently in Surrealism's visual vocabulary, notably in Bellmer's *Poupée* sculptures and photographs and Atget's Avenue des Gobelins photos, as well as in the installation of an avenue of dress mannequins at the International Exhibition of 1938. As Jentsch suggested, their uncanny nature stems, at least in part, from the fact that they have a familiar, recognizable and relatable human form, and yet strangely, they are insentient and inorganic.

Nonhuman animals are precisely the corollary: they are living, in many instances demonstrably sentient beings, and yet their form is unhuman. Here, a cross-pollination of the writings of Darwin and the ideas of Freud enables a model of what might be called a "Darwinian uncanny." Indeed, while the bodies of horses, sparrows, and lizards are easily distinguished from those of a human body, they are discernibly similar and clearly related nevertheless. Mammals, for example, but certainly not only mammals, have heads comprised of bilaterally symmetrical faces with two ears, two eyes, a nose with two nostrils, and a horizontal mouth lined with two rows of teeth. They also have bilaterally symmetrical bodies with four limbs, a heart and two lungs. In Darwin's own words:

> It is notorious that man is constructed on the same general type or model as other mammals. All the bones in his skeleton can be compared with corresponding bones in a monkey, bat, or seal. So it is with his muscles, nerves, blood-vessels, and internal viscera. The brain, the most important of all the organs, follows the same law, as shown by Huxley and other anatomists.[5]

For Darwin, this was proof of descent from a common ancestor: "On any other view, the similarity of pattern between the hand of the man or monkey, the foot of a horse, the flipper of a seal, the wing of a bat, etc., is utterly inexplicable."[6]

Derrida has written of the "right ... man ... attribute[s] to himself" to "refuse the animals such and such a power (speech, reason, experience of death, mourning, culture, institutions, technics, clothing, lying, pretense of pretense, covering of tracks, gift, laughter, crying, respect, etc.—the list is necessarily without limit, and the most powerful philosophical tradition in which we live has refused the 'animal' *all of that*)."[7] Darwin, however, had proposed, and was quite insistent upon the idea that the human resemblance to other animals is not merely anatomical, writing that

> All [the higher animals] have the same senses, intuitions, and sensations–similar passions, affections, and emotions, even the more complex ones, such as jealousy, suspicion, emulation, gratitude, and magnanimity; they practice deceit and are revengeful; they are sometimes susceptible to ridicule, and even have a sense of humor; they feel wonder and curiosity; they possess the same faculties of imitation, attention, deliberation, choice, memory, imagination, the association of ideas, and reason, though in very different degrees.[8]

Darwin's insistence upon not only the anatomical but also the emotional, behavioral, and cognitive relatedness of humans to other animals was detailed in *The Descent of Man*, published in 1871. Perhaps tellingly, however, *On the Origin of Species* of 1859, barely mentions human evolution in an otherwise laborious, if not entirely exhaustive, discussion of the mechanism of natural selection. What may be significant about the *Origin*'s rather conspicuous omission is that it suggests that Darwin regarded the critique of human uniqueness as likely to generate an even greater backlash than the scientific thwarting of creationist literalism (though the special place granted to Adam in the Garden of Eden links the two perhaps inextricably). Precisely for its decisive and inevitable critique of the metanarrative of anthropocentrism, Freud would describe Darwinian theory as, "a blow to human narcissism."[9] Derrida would affirm this conclusion: "this whole anthropocentric reinstitution of the superiority of the human order over the animal order … seems … to testify to the panic Freud spoke of: the wounded reaction not to humanity's *first* trauma, the Copernican (the earth revolves around the sun), nor its *third* trauma, the Freudian (the decentering of consciousness under the gaze of the unconscious), but rather to its *second* trauma, the Darwinian."[10]

Freud's regarding of psychoanalysis as another such blow, however, further reveals his own indebtedness to Darwin. Darwin believed that modern species continue to bear evidence of their own evolution, and he suggested, not unproblematically, that such evidence might at times reveal itself in what he called, "reversion." (For Darwin, this evidence might include such diverse traits as copious body hair, unusually pointed canine teeth, "idiocy," and "the delight shown by almost all boys in climbing trees."[11]) Following Darwin, Freud imagined that our minds, too, retain "animal" traces of our evolutionary past. Though Darwinian theory certainly does not mean—as its opponents so often and so spuriously suggest—that humans are descended from monkeys (or elephants, snails, toucans, or frogs), it does mean is that we share with these other animals a common ancestor, one long since become extinct. Coupling a Freudian point of view, with a Darwinian one, then, we might conclude that these other animals are indeed therefore alien—they are Other—at the same time that they are family.

Ultimately, Freud hypothesized that the disquieting effect of the uncanny stems from the arousal of either a repressed infantile desire, or from animistic traces retained from a precivilized, and perhaps even a prehuman past. In both cases, that of the infant, and that of the prehuman, the memory is received as a threat to the ego, to the image of the self as both fully developed and fully evolved. Both are "regressions" that the ego would prefer to avoid. The element of the unfamiliar, then, so closely associated with the uncanny might be better described as the unpleasantly familiar, as one feels cognitive dissonance in experiencing a kind of memory that one feels compelled to disavow. In the instance of dolls, for examples, Freud suggested that their uncanniness is not a matter of simple uncertainty over whether they are animate, as Jentsch had claimed, but rather that it is a result of adult discomfort over the memory of having once believed as children that our dolls in fact were or could become animate: "children do not distinguish at all sharply between living and inanimate objects, and … they are especially fond of treating their dolls like live people … the idea of a 'living doll' excites no fear at all; children have no fear of their dolls coming to life, they may even desire it. The source of uncanny feelings would not, therefore, be an infantile fear in this case, but rather an infantile wish or even merely an infantile belief."[12] Freud would similarly claim (in a separate essay) that, "A child can see no difference between his own nature and that of animals. He is not astonished at animals thinking and talking in fairy-tales; … Not until he is grown up does he become so far estranged from animals as to use their names in vilification of human beings."[13]

Unlike that of Freud, who was frequently cited overtly in the texts of Surrealism, Darwin's role in Surrealism was largely implicit. It is evident in his influence on figures such as Freud and Alfred Jarry, in the Surrealists' frequent adoption of natural history imagery, and, perhaps indirectly, in their readiness to undermine the hierarchies of the humanist and Judeo-Christian traditions. Though they dedicated themselves more explicitly to protesting colonialism, the spread of fascism, and other very human concerns, many Surrealists in a variety of ways critiqued the presumptions of anthropocentrism, including those who explored the potential of "the animal" to function as an uncanny displacement of human subjectivity and its presumptions of a unique and rational ego, the very anthropocentrism critiqued by Derrida.

If both mannequin-like objects and nonhuman animals can indeed be taken as embodiments of the uncanny, a painting by René Magritte appears to make the link explicit (Figure 10.1). *Deep Waters* depicts what at first glance appears to be a woman standing alongside a large bird. The latter is perched such that its eyes are level with those of the woman. The female figure is dressed in modern clothing, but her monochromatic head is that of a classical marble statue. Her overcoat and gloves conceal the rest of the body from view, making it impossible to determine whether she is entirely marble, or only marble from the neck up (they also serve as a reminder of the uncanny nature of other body-shaped but inanimate objects including gloves and shoes). The bird meanwhile is of indeterminate species; its size suggests a bird of prey, but it lacks the hooked beak of a hawk or a falcon. Indeed, its unthreatening facial features suggest a much more "*heimlich*" species. Each figure, then, is a paradox unto itself, and yet when combined they become perfect compliments: one is human but not live, and the other live but not human. The relationship is further reiterated visually in the painting's color

Figure 10.1 René Magritte, *Deep Waters*, 1941.

scheme. The woman's charcoal gray overcoat effectively matches the bird's dark tail and wing feathers. Its brown torso plumage is echoed in the woman's gloves, while the suggestion of her coral blouse, just visible beneath the collar of her coat, pairs visually with the similar color of the bird's talons. They are two of a kind and yet two different kinds simultaneously.

The two figures also share a relationship to the gaze of the viewer that is perhaps both contradictory and complimentary. The human head faces us, depicted frontally but with averted eyes. The bird, however, is fully in profile, but its open eye faces us laterally. It seems to look away and to see us at the same time. It is also true that while we may feel intuitively responsive to "human" head, it isn't really human at all, and even with its eyes wide open, it would never see us back. (Magritte also played with the ambiguity of this very marble head in a painting called *Memory*, in which the here disembodied head impossibly appears to bleed as if from a wound.)

Thus far, I have discussed "the animal" as a kind of corollary to automatons. As Derrida has noted, however, Descartes theorized that nonhuman animals were more directly akin to automatons. For Descartes, what they shared was an involuntary performance of actions without any ability to *respond*; they were devoid, in other words, of the sentience necessary to assess a situation and to originate an appropriate reaction. Descartes described this response as primarily—though perhaps not exclusively—linguistic. Derrida has annotated the following passage from Descartes:

> men born deaf and dumb, and thus deprived of their speech-organs as much as the beasts or even more so, normally invent their own signs to make themselves understood by those who … have the time to learn their language. This *witnesses* [my italics] not merely to the fact that the beasts have less reason than men, but that they have no reason at all. … It would be *incredible* [my italics again] that a superior specimen of the monkey or parrot species should not be able to speak as well as the stupidest child … if their souls were not completely different in nature from ours. And we must not confuse speech with the natural movements which *witness to* [my italics again] passions and which can be imitated by machines as well as by animals.[14]

Indeed, Descartes concluded that a person would be unable to distinguish between a live animal and an automaton of that animal provided that the latter were a perfect physical replica. He concluded, however, by suggesting that this same person would have no difficulty distinguishing between a live human and a human automaton for the following two reasons (this passage, too, includes Derrida's annotations):

> one is that never, unless it be by chance, do these automatons *respond* [my italics], either with words or even with signs, concerning what is *asked* [my italics] of them; and the other that although the movements that they make are often more regular and more certain then those of the wisest men, they nevertheless *lack* [my italics] several things that they should do in order to imitate us, more than would the most senseless of men.[15]

Long before Derrida, Voltaire had already expressed doubts about Descartes's formulation, stressing in almost Darwinian terms our resemblance to other animals. He wrote in the entry on "Animals" ("*Bêtes*") for the *Philosophical Dictionary*:

> There are barbarians who seize this dog, who so greatly surpasses man in fidelity and friendship, and nail him down to a table and dissect him alive, to show you the mesaraic veins! You discover in him *all the same organs of feeling as in yourself.* Answer me, mechanist, has nature arranged all the springs of feeling in this animal *to the end that he might not feel*?[16]

Nevertheless, as Peter Singer has remarked, "in the philosophy of Descartes the Christian doctrine that animals do not have immortal souls has the extraordinary consequence that they do not have consciousness either."[17]

Though the Surrealists did not seek to systematically critique the humanist presumptions of the Cartesian model of nonhuman animals as living automatons, one text from *Minotaure* (no. 5), however, comes very close to doing so. Roger Caillois's 1934 essay, "The Praying Mantis: From Biology to Psychoanalysis" begins by describing popular mantis-related beliefs and customs throughout the world, including one which maintains that, "when a mantis is asked for directions by children who are lost, it shows them the way by pointing its finger—and rarely, if ever, does it mislead them."[18]

Caillois did not similarly invoke the language of the *uncanny*, but his description of the mantis is markedly consistent with the terms employed in this discussion. Taking cues from Salvador Dalí's 1933 essay "The Tragic Myth of Millet's Angelus," he claimed, for example, that, "mankind has been highly struck by this insect. No doubt this results from some obscure sense of identification, encouraged by the mantis's remarkably anthropomorphic appearance."[19] Caillois, however, developed his own vocabulary to describe the effect of humanity's recognition of itself in another animal. He referred to such consonances as "objective ideograms," which, "*concretely realize the lyrical and passional virtualities of the mind in the outside world.*"[20] In this way, he merged the study of biology and psychology, as in his description of, "half-physiological, half-psychological factors such as man's inclination to be interested in, or even to identify with, anything whose external configuration suggests his own body."[21]

He also cited the work of Léon Binet, "a professor of physiology at the Faculté de Médicine in Paris," who had claimed that, "This insect really seems to be a machine with highly advanced parts, which can operate automatically." Caillois, however, interpreted this very Cartesian remark psychoanalytically as a projection on the part of Binet:

> Indeed, it strikes me that likening the mantis to an automaton (to a female android, given the latter's anthropomorphism) reflects the same emotional theme, if (as I have every reason to believe) the notion of an artificial, mechanical, inanimate, and unconscious machine-woman—incommensurate with man and all other living creatures—does stem in some way from a specific view of the relations between love and death and, in particular, from an ambivalent premonition of encountering one within the other.[22]

Caillois, however, acknowledged that the mantis's "jointed rigidity ... recalls a coat of armor of an automaton," adding that

> it is a fact that there are very few reactions that the mantis cannot perform in a decapitated state—that is, without any center of representation or of voluntary activity. In this condition, it can walk ... engage in mating; lay eggs ... and (this is truly frightening) lapse into feigned *rigor mortis* in the face of danger. ... I am deliberately expressing myself in a roundabout way as it is so difficult, I think,

> both for language to express and for the mind to grasp that the mantis, when dead, should be capable of simulating death.[23]

Freud would likely affirm that that mantis's simulation of death while dead is indeed uncanny, but there are additional implications to Caillois's description of the "automatic" behavior of the decapitated mantis. For example, in "Animal Mimicry and Psychasthenia," he noted that humans, too, have automatic responses to particular stimuli, such as, "the human phenomenon termed *gooseflesh*, which especially occurs at times of great fright. It has persisted, even though the atrophy of the pilose system has made it obsolete."[24]

Caillois would even come to the very un-Cartesian conclusion that, "both these insects and mankind are part of one and the same nature," and he would add, "I do not exclude the possibility of invoking the insects to explain, if need be, people's behavior in certain situations. *For we must realize that man is a unique case only in his own eyes*, and that this study is actually nothing but comparative biology."[25] Though his theories may be scientifically unverifiable—perhaps at best—they do indeed provide a specific, albeit theoretical model for imagining a very Darwinian experience of the uncanny, that calls out the presumptions of anthropocentrism.

In his critique of Descartes, Derrida has written that the former "appeals to a man who sees an animal that doesn't see him,"[26] and he reminds us that this view was based on a *judgment* by Descartes. Derrida defined himself conversely, as a man who sees an animal who *does* see him: "nothing will have ever given me more food for thinking through this absolute alterity of the neighbor or of the next(-door) than these moments when I see myself seen naked under the gaze of a cat."[27]

It was this very scene that became the launching pad for Derrida's critique of speciesism in Western philosophy, and he developed its implications at length. Specifically, he described in it a feeling of being overcome with shame at these moments, followed by a feeling of being ashamed of his own shame. At this, he asked himself: "Ashamed of what and before whom?"[28] The encounter instigated a crisis for Derrida (who initiated this text with the very question that Breton used to open *Nadja*: "Who am I?"[29]) Derrida quickly identifies the source of his shame; he is, "Ashamed of being as naked as a beast."[30] In his nudity, he is like the cat, whom Derrida insists is a real cat and not "the *figure* of a cat,"[31] but in his shame he is different: "It is generally thought, although none of the philosophers I am about to examine actually mentions it, that the property unique to animals, what in the last instance distinguishes them from man, is their being naked without knowing it."[32] This generates a paradox in that the very concept of nudity is meaningless to a nonhuman animal: "naked without knowing it, animals would not be, in truth, naked. They wouldn't be naked because they are naked."[33]

That she is unclothed furthermore distances her from the spectacle of anthropomorphization that became popular in Europe in the nineteenth century (and remains a phenomenon in the internet era).[34] She is not like the monkeys, dogs, and cats dressed in jackets or hats in a manner that acknowledges their likeness to humans only to package it as a humorous amusement in which the nonhuman other is the object of a one-way human gaze. Ultimately, what is most important about this scene, then, is the question of what it means to be seen by a nonhuman animal, not merely to be perceived, but to be seen, to recognize in the animal a seer, a subject with the potential therefore to decenter the master narrative that is anthropocentrism, and the personal (and cultural) identities that humans construct for themselves around this master narrative. Though the domesticated house cat might seem to belong to the realm of the *Heimlich*, as described by Sanders, in *seeing* him, in claiming ownership of her own gaze and thereby enacting a subjectivity that renders him *her* Other, she becomes *unheimlich*.

As I have indicated, the question of "the animal" was not explicitly integral to the manifestoes or other didactic texts of historical Surrealism. Breton did not, as did Derrida, systematically challenge Descartes' presumptions or other manifestations of speciesism in Western religion and philosophy, although one passage from *Arcanum 17* pointedly described what Freud would call "human narcissism":

> At the top of the list of … errors that remain the most detrimental stands the idea that the universe only has intelligible meaning for mankind, and that it has none, for instance, for animals. Man prides himself on being the chosen one of creation. Everything that evolution has been able to reveal to him about his origins and general biological necessities … remains in effect a dead letter. He persists in seeing and acting as though these revelations, so crushing to his pride, never occurred."[35]

There are also in Surrealists works instances of encounters between humans and other animals that are perhaps pointedly reminiscent of the naked Derrida's encounter with his cat. These similarly evoke a sense of the Darwinian uncanny, as a potentially decentering confrontation with an Other across species lines.

The Civilizer, a painting by René Magritte depicts a frontal, centered, head-and-shoulders image of a face with eyes directly engaging the gaze of the viewer (Figure 10.2). This is to say that it is a portrait, but it is a portrait of Magritte's dog (as with Derrida, this is the artist's own domestic companion). The genre of portraiture is one that had been historically reserved, as if by definition, for human subjects. Through the Renaissance, the convention of the frontal portrait was even reserved for holy or royal subjects. The

Figure 10.2 René Magritte, *The Civilizer*, 1946.

European academies recognized, though they did not highly esteem, a genre known as "animal painting," but the conventions of this genre typically dictated a full body image seen at a remove, as if the animal were part of a taxidermied display, such as one finds in the eighteenth-century paintings of George Stubbs. Magritte's image, however, does not adhere to these conventions. Alex Potts has noted many changing conventions in representations of nonhuman animals, and the later part of the nineteenth-century gave rise to an era in which proximity to, "pets and zoo animals … led to their being given names and being represented in portraits," as in many works by Rosa Bonheur and Thomas Landseer.[36] Nevertheless, Magritte's painting remains exceptional in the prioritization of the subject's face and the directness of the subject's gaze. Indeed, the latter is perhaps even reminiscent of the infamously confrontational gaze of Manet's *Olympia* (1863), and it may or may not be coincidental that *The Civilizer* was painted during a juncture in Magritte's career when he attempted to infuse Surrealism with Impressionist imagery.[37] This interest is furthermore evident in the painting's soft color palette and the looser, more painterly technique employed by the artist whose better known works are typically hard-edged and stylistically deadpan. (Magritte also painted an explicit reference to Manet's *Olympia* in 1948. In this instance, however, the nude woman's gaze is directed toward the conch shell upon her abdomen, and she does not, as in Manet's famous precedent directly confront the gaze of the viewer.)

The background landscape of *The Civilizer* is also ambiguous. It appears to be a dense forest, but it is a forest cut short by a broad ring of yellow grass that separates the subject from the trees. The distance between the dog and the forest, however, is reminiscent of a distorting wide-angle photograph, and it heightens the sense of the dog's immanence before the viewer. In their ability to take verbal commands, dogs are perhaps the canniest of all nonhuman animals, but they are canny precisely because they seem to "know their place," and to be willing to defer to humans. Like Derrida's cat, however, this dog appears to assume ownership of its own gaze. In the process, even the animal most familiar retains, or perhaps reclaims, an ability to be experienced as strange.

Of course the painting's title, *The Civilizer*, is also worth noting. While Magritte often employed pointedly enigmatic titles, the inevitable assumption here is that it is the dog who is the Civilizer. Eighteenth-century theories of child development had suggested that children might in effect civilize themselves through the process of domesticating an animal such as a squirrel, but the notion of a nonhuman animal—even the *heimlich* dog—*being* the civilizer may seem unsettling in its reversal of expected roles. It is yet another way in which this *heimlich* and familiar animal may well be experienced as strange and *unheimlich.* A second version of the image places the dog in front of a classical marble temple, itself apparently a bastion of civilization, and yet it is still the dog who seems the civilizing agent of the title.

The films of Luis Buñuel also provide examples of uncanny interspecies encounters notable for the intensity of the nonhuman gaze. *El Bruto* (1952) dates from the period of his Mexican "melodramas," and indeed it conforms to many of the genre's conventions, but with the characteristic inclusion of numerous "Buñuelian" touches. Paloma, the film's female lead, has a name meaning *dove*. Like the stork, this is an animal admired for its beauty and gentility; it is nonthreatening, *heimlich* even. Conversely, the man called "brute" is likened to a "beast," a wild, *unheimlich* animal, and he is scorned as such. In a reversal of expectations, however, it is Bruto who becomes the sympathetic—if not entirely innocent—victim of Paloma's machinations.

After Bruto is killed in a shoot-out with police that has been instigated by Paloma, the final scene follows her as she makes her exit from the courtyard. She leaves behind the unruly mass of police and onlookers as she enters a barn with the intent only of passing through it. She finds, however, that she is not alone, for there is a rooster in the barn, and he is looking at her (Figure 10.3). The bird stands on a ledge, where his gaze meets hers at her eye level. She slows her pace, and her eyes widen. She begins to pass the rooster, but turns her head to keep him fixed in her gaze. She utters no speech—for what would she say to him?—but it is clear from her expression that she has found the rooster's return of her own gaze disconcerting. There are no other animals in the barn—human or otherwise—to detract from the intensity of this encounter. In a point-of-view shot, we see what Paloma sees; a close-up of the rooster reveals his blank stare. He does not "speak"; he merely regards

Figure 10.3 Lius Buñuel, still from *El Bruto* (1952).

her, as if her motives, her intentions, and even her actions are no more consequential to him than his are to her.

It may be that in seeing the rooster see her, she comes to see herself more deeply, as if the rooster were an external embodiment of her own conscience, but she shows no outward sign of repenting. Indeed, as John Berger has written, "No animal confirms man, either positively or negatively."[38] Ultimately, the scene is ambiguous, and it would not be unlike Buñuel for it to be so purposefully. Indeed, it was likely an attempt on Buñuel's part to end the commercially made melodrama on a strangely poetic note. But what makes it strange—and what makes it poetic—is precisely the encounter of two beings who are different from one another, in their lack of a common language, but alike in their shared possession of a gaze capable of being fixed upon another.

In this scene, the viewer observes a confrontation between two filmic characters, thus assuming the role of the "third person," (note the terminology *first*, *second*, and *third person*). In *The Phantom of Liberty* (1974), however, it is the viewers who themselves experience the encounter with a bird, one that reads on many levels as uncanny. In one of the film's tangentially linked episodes, a man sits awake in his bed at night only to find a stream of unlikely visitors pass through his room. The first is here, too, a rooster, the second a woman carrying a candle, and the third is a mail carrier on a bicycle. The fourth and final interloper is an ostrich. We are tempted to interpret these nocturnal visits as dreams, but Buñuel foils this conclusion by indicating that the man is still able to produce the physical letter left by the mail carrier the next day.

Figure 10.4 Luis Buñuel, still from *The Phantom of Liberty* (1974).

There are moments in the ostrich sequence, albeit brief ones, in which the bird looks directly at—or at least in the direction of—the camera. This kind of engagement is usually avoided in narrative cinema, for it violates the "fourth wall," and destroys the illusion of self-containment. In the process, it reminds the viewer that they are an embodied spectator, who has now in effect been caught looking. In this instance, the one who has caught us is an ostrich. The camera lingers on the face of the bird in close-up, a shot typically reserved for human faces (Figure 10.4).

Derrida has criticized Levinas for articulating an ethic of the Other that neglects to consider what Carrie Rohman has called the "radical alterity" of the animal Other.[39] For Levinas, it is the image of the face that defines this ethic. Derrida has quoted Levinas' reply to the question, "Does the animal have a face?": "I cannot say at what moment you have the right to be called 'face.' The human face is completely different and only afterwards do we discover the face of an animal. I don't know if a snake has a face. I can't answer that question.'"[40] In this sequence Buñuel shows us the ostrich's face. Though this may seem a much more literal use of the term than anything ever intended by Levinas, the scene at the very least reminds us that Buñuel's treatment of the ostrich defies cinematic conventions by filming the bird almost as if it were another human character.

The *almost*, however, is significant here because we know that unlike the other actors in the film, the ostrich is not acting. It is by contrast being. It apparently has taken some direction in entering the space and pausing, but in the close-up shot, we see that this bird is taking in and assessing its surroundings. Its gestures seem indicative of curiosity, behaving as it might in any unfamiliar space. It becomes difficult not to wonder during this silent, drawn out close-up how the ostrich is experiencing this, what it makes of the sets and the actors, not knowing that they are sets and actors. We might even imagine the lights, the cameras, and Buñuel in the director's chair. The ostrich does not perceive this "scene" as illusion, cinematic or otherwise, and in seeing the ostrich see the trees instead of the forest, the illusion may be compromised, if only momentarily, for the viewer, too. This breakdown of cinematic codes might also be experienced as uncanny, as a kind of tear in the fabric of the Symbolic order, a "reversion" to the real that comes from assuming an identification with an ostrich that looks back at the camera.

Buñuel reiterates the significance of this moment by concluding the film with another protracted shot of an ostrich. This occurs during the "riot at the zoo." Here, too, Buñuel is willfully enigmatic, providing no details as to the motivation of the rioters. The suggestion that it may be an act of animal liberation is perhaps inevitable, but the film neither confirms nor denies this possibility. We learn only through a brief conversation between two police officers, that they have foreknowledge of the evidently planned event, and that they have intentions to suppress it. Their remarks, however, indicate that concern for the animals is not a priority:

> "The main thing is to keep them from reaching the cages."
> "The locks have been carefully checked."
> "And anyway, if some of the animals are killed, tough luck."
> "Our men are more valuable than a zebra."

When the police arrive at the zoo, the animals therein become the sole focus of the camera. Shots of a rhinoceros, a seal, a hippopotamus, an elephant, a vulture, a bear, and a lion pacing behind bars are set to the sounds of the riot including gunshots and human screams. These animals, most of whom do not look into the camera, seem conspicuously both the object of a human gaze—they are after all "exhibits" in a zoo—and the agents through which we as viewers experience the riot. Without verbal exchanges or a visual representation of its human protagonists, the actions lose their ability to function as narrative and become only sounds. It is thus for both the animals, who have no context for comprehending the human conflict, and the human spectators of the film.

The final shots linger on the face of an ostrich. Upright, two-legged, and human scale, the ostrich may be especially uncanny in its structural resemblance to a human animal, and indeed it is an animal that might look a person in the eye while standing face to face. As in the previous ostrich close-up, however, Buñuel gives us ample time to contemplate the face of a creature both markedly like us and also decisively Other.

Berger has noted that an animal, "does not reserve a special look for man," and he has remarked that "the same animal may well look at other species in the same way." Seeing oneself through the eyes of a nonhuman animal, then, may ultimately be to see oneself as animal. Berger elaborates: "The animal scrutinises [man] across a narrow abyss of non-comprehension. This is why the man can surprise the animal. Yet the animal–even if domesticated–can also surprise the man."[41] Such surprises, however, may be unwelcome in their undermining of a humanist sense of the self as not-animal, what Freud called, "human narcissism." In their embrace of the uncanny, however, numerous Surrealists explored this very potential, and the potential for humans to see in other animals that which is both like and unlike.

In this way they anticipated Derrida but in an asymmetrical way. Nonhuman animals were a vehicle through which they explored the broader experience of the uncanny, while uncanniness was but one facet of Derrida's elaborated discussions of nonhuman animals (though indeed he describes the uncanniness of his cat's gaze without naming it as such). In both instances, however, the intersection occurs, as if inevitably, at the point at which "the animal" is perceived not as a seen object but as a seeing subject, the very point at which we become other to the other.

NOTES

1. Sigmund Freud, "The Uncanny" (1919), in *The Complete Psychological Works of Sigmund Freud,* volume XVII (London: Hogarth Press, 1953), 222.
2. Ibid., 245.
3. Ibid., 226.
4. Jacques Derrida, *The Beast and the Sovereign* (Chicago: University of Chicago Press, 2009), 187.
5. Charles Darwin, *The Descent of Man and Selection in Relation to Sex* (New York: American Publishing Corp, 1874), 22.
6. Ibid., 40.
7. Jacques Derrida, *The Animal The Animal that Therefore I Am* (New York: Fordham University Press), 2008, 135.
8. Darwin, *Descent of Man*, 94–95.
9. Sigmund Freud, "A Difficulty in the Path of Psycho-Analysis" (1917), in *The Complete Psychological Works of Sigmund Freud*, volume XVII (1917–19), 141.
10. Derrida, *The Animal That Therefore I Am*, 136.
11. Darwin, *The Descent of Man*, 52.
12. Freud, "The Uncanny," 233.
13. Freud, "A Difficulty in the Path of Psychoanalysis," 140.
14. Derrida, *The Animal That Therefore I Am*, 76.
15. Ibid., 82.
16. Voltaire, *Philosophical Dictionary*, Hanover Historical Texts Project, https://history.hanover.edu/texts/voltaire/volanima.html.
17. Peter Singer, *Animal Liberation* (New York: HarperCollins, 1975), 200.

18. Roger Caillois, "The Praying Mantis: From Biology to Psychoanalysis," in *The Edge of Surrealism: A Roger Callois Reader*, edited by Claudine Frank (Durham: Duke University Press, 2003), 70.
19. Ibid., 73.
20. Ibid., 80.
21. Ibid., 81.
22. Ibid., 78–79.
23. Ibid., 79.
24. Roger Caillois, "Animal Mimicry and Psychasthenia," in *The Edge of Surrealism*, 92n.
25. Ibid., 69n, emphasis added.
26. Derrida, *The Animal That Therefore I Am,* 82.
27. Ibid., 11.
28. Ibid., 4.
29. André Breton, *Nadja* (New York: Grove Press, 1960), 3.
30. Derrida, *The Animal That Therefore I Am,* 4.
31. Ibid., 6.
32. Ibid., 4–5.
33. Ibid., 5.
34. See Diana Donald, "'A Mind and Conscience Akin to Our Own': Darwin's Theory of Expression and the Depiction of Animals in Nineteenth-century Britain," in Diana Donald and Jane Munro, eds. *Endless Forms: Charles Darwin, Natural Science and the Visual Arts.* New Haven: Yale University Press, 2009), 195.
35. André Breton, *Arcanum 17* (Los Angeles: Sun and Moon Press, 1994), 44.
36. Julia Voss, "Monkeys, Apes and Evolutionary Theory: From Human Descent to King Kong," in *Endless Forms: Charles Darwin, Natural Science and the Visual Arts* (New Haven: Yale University Press, 2009) 219.
37. See Siegfried Gohr, *Magritte: Attempting the Impossible* (New York: Distributed Art Publishers, Inc., 2009), 191–94.
38. John Berger, "Why Look at Animals?" 5.
39. Carrie Rohman, *Stalking the Subject: Modernism and the Animal* (New York: Columbia University Press, 2009), 17.
40. Derrida, *The Animal That Therefore I Am,* 107.
41. John Berger, "Why Look at Animals?" 4–5.

WORKS CITED

Berger, John. "Why Look at Animals?" in *About Looking.* New York: Pantheon, 1980.

Breton, André. *Arcanum 17.* Los Angeles, CA: Sun and Moon Press, 1994.

———. *Nadja.* New York, NY: Grove Press, 1960.

Caillois, Roger. "Animal Mimicry and Psychasthenia," in *The Edge of Surrealism: A Roger Callois Reader*, edited by Claudine Frank, 89–103. Durham, NC: Duke University Press, 2003.

———. "The Praying Mantis: From Biology to Psychoanalysis," in *The Edge of Surrealism: A Roger Callois Reader*, edited by Claudine Frank, 66–81.Durham, NC: Duke University Press, 2003.

Dalí, Salvador. "The Tragic Myth of Millet's Angelus," (1933) excerpted in *The Collected Writings of Salvador Dalí*. Cambridge: Cambridge University Press, 1998.

Darwin, Charles. *The Descent of Man; and Selection in Relation to Sex*. New York, NY: American Publishing Corp, 1874.

Derrida, Jacques. *The Animal that Therefore I Am*. New York, NY: Fordham University Press, 2008.

———. *The Beast and the Sovereign*. Chicago, IL: University of Chicago Press, 2009.

Donald, Diana. "'A Mind and Conscience Akin to Our Own': Darwin's Theory of Expression and the Depiction of Animals in Nineteenth-century Britain" in *Endless Forms: Charles Darwin, Natural Science and the Visual Arts*, edited Diana Donald and Jane Munro, 195–214.New Haven, CT: Yale University Press, 2009.

Freud, Sigmund. "A Difficulty in the Path of Psycho-Analysis," (1917) in *The Complete Psychological Works of Sigmund Freud*, volume XVII (1917–19), 135–144. London: Hogarth Press, 1953.

———. "The 'Uncanny,'" (1919), in *The Complete Psychological Works of Sigmund Freud,* volume XVII, 217–252.

Gohr, Siegfried. *Magritte: Attempting the Impossible*. New York, NY: Distributed Art Publishers, Inc., 2009.

Potts, Alex. "Natural Order and the Call of the Wild."*Oxford Art Journal*, volume 13, No. 1 (1990), pp. 12–33.

Rohman, Carrie. *Stalking the Subject: Modernism and the Animal*. New York, NY: Columbia UP, 2009.

Singer, Peter. *Animal Liberation*. New York, NY: HarperCollins, 1975.

Voltaire. *Philosophical Dictionary*. Hanover Historical Texts Project. Accessed April 29, 2016. https://history.hanover.edu/texts/voltaire/volanima.html.

Voss, Julia. "Monkeys, Apes and Evolutionary Theory: From Human Descent to King Kong," in *Endless Forms: Charles Darwin, Natural Science and the Visual Arts,* edited by Diana Donald and Jane Munro, 215–236. New Haven, CT: Yale University Press, 2009.

Chapter 11

Becoming-Animal and the Two Meanings of Animality

A Derridean Reading of Black Swan

Rodolfo Piskorski

Deleuze and Guattari open their chapter "1730: Becoming-Intense, Becoming-Animal, Becoming-Imperceptible" in *A Thousand Plateaus* with an account of a film titled *Willard* (1971, Daniel Mann, but also remade in 2003 by Glenn Morgan). For them it is a film about a man's becoming-rat.[1] I propose that Deleuze and Guattari find in *Willard* a paradigmatic structure of narrative and themes that was already present in another famous story about becoming-animal: the 1875–6 Tchaikovsky ballet *Swan Lake*. These texts present alluring animal becomings which go beyond a symbol of escape from restrictive human societies. The lines of flight, to use their term, engendered by these becomings-animal are not—at least not mainly—constructed on the basis of negation of human sociality, which would consist in one of the two vulgar meanings of animality. In the ballet, the Prince's sexuality (curtailed in its free expression by the parental indictment to procreate and contribute to the Oedipal structure) is rerouted towards dancing with swans beside the mysterious lake he encounters in the woods.

In order to analyze this animal becoming, this chapter will read Darren Aronofsky's 2010 film *Black Swan*[2] as a reworking of *Swan Lake* which in fact exposes the becoming-animal of the ballet for its disruptive qualities. By putting the ballerina Nina Sayers center stage, the film conflates the ballet's Prince and the swan-maiden Odette in one character which is both transformed into a swan and attracted to its non-Oedipal possibilities. However, if a certain attention to Nina as a character in this densely psychoanalytic film is warranted, I propose an analysis of this becoming-swan that reads it for its textual effects in the film. Or rather, for its extratextual effects.

I would like to explore the possibility that animality is not completely *saturated* by textuality. I propose that what I call *arche-animality*[3] somehow escapes signifying projects and practices in a way which is similar to

Deleuze and Guattari's becoming-animal's lines of flight. This escape, unlike the overdetermined field of (animal) embodiment, is not constituted by that from which it escapes and does not, ultimately, mean anything. I argue that *Black Swan* as a text concerns itself entirely with the mapping and manipulation of the dancer's body as the site of tension between different accounts of animality (how does the human "inhabit" his or her animal body, and what can this animal *mean*?). I shall show that the film stages the two standard philosophical conceptions of animality in regards to corporeality: materialism and transcendence. A materialist account of animal being emphasizes that animal embodiment completely escapes conceptuality in a way that places animals beyond the grasp of signification. On the other hand, a Cartesian transcendentalist approach would stress the fact that not only are animals like us since they too have minds encased in bodies, but that their bodies—and the *meaning* of their bodies—are subordinate to (human) consciousness.

However, both of these approaches are unsatisfactory. The latter, Cartesian line of thinking has more than fallen out of fashion as a productive use of thinking about animality, while the former, materialistic account is naïve in thinking that matter is not constituted *as* the other of textuality *by* textuality itself. This selfsame refusal to accept both sides of a dichotomy is the gesture performed by Derrida throughout his career, from *Of Grammatology* to *The Animal That Therefore I Am*. In the former, Derrida argues that both writing and speech are only the dissimulation of a more originary element, which he calls by means of the *paleonym* (old name) "arche-writing." In the latter, as I shall argue, he proposes the word *l'animot* as the source of the distinction between the singular *l'animal* and the general *les animaux*.

Leonard Lawlor identifies paleonymy with the "second phase" of deconstruction. The second phase "reinscribes the previously inferior term as the 'origin' or 'resource' of the hierarchy itself," so that this term "becomes what Derrida calls an 'old name' or a 'paleonym'."[4] He explains the Derridean paleonym "pharmakon" in terms that also apply to *Of Grammatology*'s arche-writing, since many of Derrida's neologisms can stand for each other in a chain of signification. Lawlor argues that the pharmakon "refers to the resource called the *logos*, language, but language prior to the division between living voice and dead writing." It is "the 'indissociability' of the signifier and signified concept."[5] He continues:

> The pharmakon then refers to language prior to the decision to separate and make determinate the signified concept or form from the sensible body of language; it refers to language prior to the decision to value the form more than the matter. The pharmakon is therefore prior to the decision that instituted Platonism or metaphysics. [...] The pharmakon is *not* a medium [...] in which prior pure elements come to be mixed; it is not a mixture that is second; it is also not

the simplicity of a *coincidentia oppositorum*. [...] Being prior to all oppositions and purifying separations, the pharmakon is an 'element-medium', which means that the *milieu* is itself the prior element.[6]

In *Of Grammatology*, Derrida places the subordinate term "writing" as the source of its dualism with "speech." But writing in the vulgar sense is only ever a modification of this more originary arche-writing, so that the terms of the dichotomy are produced by means of a repression of the element named by the paleonym. And the material, bodily aspects of writing—when compared to the spirituality of speech—should reveal the extent to which the paleonymy at work in arche-writing is also applicable to the issue of animal corporeality.

This will be crucial for any reading of *Black Swan* when one is asked to determine what its intertextual relationship to Tchaikovsky's ballet *means*. It is safe to say that the audience is led to believe that the madness suffered by the protagonist is caused (or at least informed) by the libretto of the ballet on which she is working, so that her becoming-swan is revealed as a suggestive balletic hyperbole of madness. More than that, however, it is clear that the plot of the ballet is particularly challenging for the character due to her own psychosexual plot. In other words, it would appear that a personal, "real-life" psychosexual conflict is replaced by a ballet *metaphor*, which is in its turn duplicated metaphorically by Nina's mad bodily transformations.

Contrary to that, I argue that the arche-animality "present" in the film (which gets repressed by the ostensive "animal" signifiers)—or its becoming-animal—disrupts the metaphorical relations between the textual strata. Becoming-swan may be a line of flight suggested by the ballet for escaping psychosexual normativity, but equally, such normality is also, to a certain extent, conditioned by the fictional libretto. The visual texture of *Black Swan*, therefore, exposes a certain *operatics* or *histrionics* of the Oedipus Complex.

Rather than doing its classical metaphorical work, animality in *Black Swan* points to a way out of the oversignification of the psyche and the dualism of animal corporeality. Psychoanalysis, as critiqued by Deleuze and Guattari, strives to map all phenomena into the Oedipal territory with which it is familiar. Becoming-animal is precisely the resistance to this territorialization. As such, it mirrors arche-animality in its independence from the stifling dialectics of meaning and representation. Traditional humanism and animal materialism are two sides of the same process and actually depend on each other for their articulation. As Derrida has shown,[7] it is ineffective to simply rely on materialism to solve the philosophical problems of metaphysics. Vulgar, commonsensical, metaphysical animal embodiment ought to be signaled as the modification of a prior, more generalized difference which metaphysics represses.

Thus, I start my discussion with a brief outline of the film's plot as evidenced by my analysis of one significant scene, along with what could be called a standard Lacanian interpretation. After that, I delve more deeply into some of the thematic strands that organize the textuality of the film and set up its investment in arche-animality. By doing that, I replicate Derrida's signature gestures and techniques when reading—a deconstructing—dichotomies. Finally, I read the disruptive qualities of this arche-animality, before connecting it both to Deleuze and Guattari's becoming-animal and the film's becoming-swan.

MIRRORED SWANS

In a crucial scene in *Black Swan*, Nina and her mother Erica are both reflected in a three-part mirror while adjusting Nina's ballet slippers, but in a cross-reflection: Nina is reflected next to her mother and Erica is reflected close to Nina.[8] What is especially striking in this cross-identification reflection is the fact that the central mirror does not reflect anything. Something in the three-part structure suggested by the mirror is missing in this moment of mother-daughter mirroring (see Figure11.1).The scene seems to inquire about the element suggested by such absence. The familial context would invite the image of a father, or another female element halfway between mother and daughter, a sort of intermediate state between girl and woman. The alignment of the central mirror with the frame of the shot, however, seems to suggest the appearance of the viewer's own reflection, or of the filmic apparatus (camera, director, etc.). The growing distance between mother and daughter that has been building up in the film prior to this scene might indicate the reason for such a gap between the two reflections, as well as the larger distance between Nina and the mirror.

Of the film's many scenes containing mirrors, this may be the one which most warrants a Lacanian reading, one which would argue that the connection

Figure 11.1 Nina and her mother Erica sit near the three-part mirror *(Black Swan)*.

between mother and daughter depicted in the film is a metaphor for Lacan's Imaginary order, which is precisely concluded by the Mirror Stage. According to that, Nina is being smothered by her mother's Imaginary identification in which Nina is the phallic signifier that compensates for Erica's failed career. Because of that, Nina seeks in the mirrors that surround her the possibility of constructing an ego of her own. But precisely because of her "madness" (which would be a result of her mother's domination), mirrors do not offer her anything other than *doubles*—mirrored images which she doesn't recognize as a simple reflection of herself. And the father figure, which by means of the Name of the Father would structure the "healthy" girl's sexuality when she enter the Symbolic order, is only marginally present due to the mother's exaggerated influence. Erica does not admit that Nina break the child-like relationship with the mother and identify with the image of a woman, sexually available to a man-father. Because of that, Nina sees her own impulses which desire precisely that as another person: her understudy or alternate, the sensuous Lily.

The dual structure of a personality that does not identify with its own double is introduced primarily by Nina's professional dilemma as the soloist in *Swan Lake*, having to dance two completely different characters. To be cast as both Swans, Nina must capture the male attention of the company director, Thomas. In this quest, she identifies with the company's prima ballerina who is about to retire, Beth, who enjoys an enviable intimacy with him. Nina wishes to impress the director, to be able to "seduce" him and catch his attention like Beth.

Such rivalry for male attention is sharpened during the auditions for the soloist in Tchaikovsky's ballet. Nina easily expresses the sweetness and vulnerability of the White Swan, but she must also show skill with the sensual choreography for the Black Swan. Thomas tells her to "show [him her] Black Swan" (*Black Swan*), but during the famous *fouettés* in Odile's Coda,[9] Nina is interrupted by Lily's bursting into the room and slamming the door, appropriately dressed all in black. Nina loses balances, and stumbles, and when she asks Thomas if she should try again, she must wait for several seconds for his negative answer while he admires Lily.

The psychoanalytical structuring of the film is thus clear: Nina wishes to be desired by Thomas, the father figure, and for that she identifies with Beth as her adult *imago*. However, the typical process of the mirror stage which would have her identify in her own body the image (even if a false one) of a coherent and mature being is problematic for Nina. Her domineering mother seems to preclude such identification, and thus Nina displaces this ego-ideal to her rival Lily. She is then visited by the "phantasm of the broken down body"—the incoherent and fragmented body of early childhood which is mitigated by self-identification in the mirror stage—when her body apparently starts to dissolve.

That both scenes described above (the one with the three-part mirror, and that of the audition) prominently feature mirrors is understandable in a film about ballet. Dancers follow their own dancing in their mirrored reflections so as to assess their skill: that is the diegetic reason for the large mirrors which are present even in Nina's apartment, where she practices. Cinematically, on the other hand, mirrors both explore the nature of the screen itself (as well as other optic apparatuses, such as lenses), and visually represent characters' relationships to themselves. Thematically, *Black Swan* revels in the latter cinematic function (as well as the diegetic one), whereas an even stronger impact may be garnered by the former function, of mirrors as textual hinges between different realms of signification. All these functions seem to coalesce around the prominent Lacanian concept of the mirror stage, which is, not surprisingly as I hope to show, conceptually formulated in relation to animality.

Lacan states that the "conception" of the mirror stage "originated in a feature of *human* behaviour illuminated by a fact of *comparative* psychology,"[10] that is, a comparison of human and animal psychologies. A human child, he describes,

> at an age when he is for a time, however short, outdone by the chimpanzee in instrumental intelligence, can nevertheless already recognize as such his own image in a mirror. [...] This act, far from exhausting itself, as in the case of the monkey, once the image has been mastered and found empty, immediately rebounds in the case of the child in a series of gestures in which he experiences in play the relation between the movements assumed in the image and the reflected environment.[11]

For *Black Swan*, Lacan's grounding of the mirror stage on species difference seems particularly crucial since it is, according to most accounts, a film about the horror of one's double when it takes an untameable animal form. In other words, the horror seems to lie on the fact that the otherness of her double, besides clearly emanating from herself, reveals the otherness of animality. In the plot, doubling is triggered by Nina's professional requirement to both perform two characters, and to outperform her rival—and later understudy—Lily, who threatens to take both roles. It certainly doesn't help that most ballet dancers shown in the film look similar. The paradoxical working of the mirror stage—where one's reflected image is, to a certain extent, both self and other—is clearly reproduced in the plot: Nina must identify with Odile in order to play her, but is constantly projecting her onto Lily, who supposedly embodies the character more easily.

Viewers are led to believe that the animal shape taken by this double is thoroughly informed by the libretto of the ballet. Because Nina's problem is tackling the Black Swan as a character (which, we learn, means locating within herself a freer sexuality than the one she is used to), she sees herself being threatened in her hallucinations by the big, black bird of *Swan Lake*, which comes to stand both for the sensual Lily and for the her own stifled sexual persona. It certainly must have occurred to Nina herself that *having her place taken* by a seductive but otherwise identical woman is precisely the fate of Odette—that would be, indeed, the psychological, or even Oedipal, reading of the animal double in the film. I propose, however, that a more thorough perusal of the signifying practices and the visual language employed by the film when exposing its plotlines can point to a different understanding of becoming-swan with regard to the suffocatingly multi-layered textuality around it. That is, one can read the animal in the text—the swan—as the product of repression of arche-animality.

Transcendanse

The premise outlined above regarding the dual role of Black and White Swan is introduced very early in the film and visually codified throughout, in a signifying saturation which recalls the over-determination of dreams. In one of the first scenes, Nina informs her mother, a retired dancer, that the company director may feature her more prominently this season, to which her mother replies: "Well, he certainly should. You've been there long enough" (*Black Swan*). The comment is clearly not well-received by Nina, who detects the implication that she is getting old and that her position is threatened by younger dancers. In the underground on the way to the theatre, Nina is reflected in the darkened window of the train, whose arrival at a station, rushing by lights and other trains, is accompanied by the rustling sounds of flapping wings, which are also visually represented in her feathery scarf. While Nina admires her own reflection and contemplates whether her place will be taken by another, the film frames this mirroring and doubling movement in animal, bird terms, foreshadowing the Black Swan as the signifier for the usurping doppelgänger.

If the glass of the train window functions as a somewhat effective reflector, the more obviously transparent glass of the door between two wagons offers what appears to be a truer mirroring. Through the door, she sees a girl in the next wagon who looks very much like her and is wearing her hair the same way, but is dressed in dark clothes. The girl indeed has *the same face*, since she is in fact being played by Natalie Portman. While staring puzzlingly at her doppelgänger, who has quickly turned her face away, Nina notices that her brushing of her hair behind her ear is mirrored by the girl. Whereas the actual reflection seen in the window is a poor, blurry image, the reality beyond the glass door not only looks but also acts the same. Thus, this initial scene blurs the essential distinction between transparent glass and reflective mirror, exterior reality and self-referential interiority. This blurring has consequences both for the thematic concerns of the film—the double is both inside and outside oneself—but also to the functioning of its visual textuality.

Replacing and doubling continue to be explored in the next scenes: Nina stops to admire the two identical season promotion posters featuring the retiring prima ballerina Beth, obviously hoping to take the latter's place, and in the female dancers' dressing room, the topic of conversation is precisely Beth's refusal to realize that she is past the age when she could attract audiences. The room is so crowded with mirrors, that the viewer can easily feel lost as to the number of girls in it and their position relative one another.

During their collective warm-up, the dancers are joined by Thomas, the company director, who summarizes the plot of *Swan Lake*:

> We all know the story. Virginal girl, pure and sweet, trapped in the body of a swan. She desires freedom, but only true love can break the spell. Her wish is nearly granted in the form of a prince. But before he can declare his love, her lustful twin, the Black Swan, tricks and seduces him. Devastated, the White Swan leaps off a cliff, killing herself, and in death finds freedom. (*Black Swan*)

He then pronounces his challenge: "Which of you can embody both swans? The white and the black." And regarding the staleness of *Swan Lake* as a standard part of the repertoire, he argues that it has "not [been done] like this. We strip it down, make it visceral and real." This new production thus "needs a new Swan Queen. A fresh face to present to the world" (*Black Swan*). In his presentation, Thomas reveals a great deal about his interpretation of the ballet. If the dancers act towards him as if he were a Prince figure, pining for his attention and approval, it is clear that he is also Rothbart,[12] willing to transform a girl into a seductive performer for a captivated audience (taken to mean the Prince and the ball guests, the ballet audience, and the wider world both inside and outside the film).[13]

Thomas's version of *Swan Lake* is, as many others, the story of Odette, and not the Prince. But more than that, it is the story of Odette's *body*, as woman, as swan, and as image to be copied by Odile. The bodily prison of swan-being registers as a familiar Cartesian indictment against the human (animal) body itself as a cage for true humanity—spiritual reason and, in this case, love. Descartes' metaphysical belief that all bodies are animal bodies—and that all animals are machinic—is literalized in the ballet in the story of a human inside an animal body. Not even Odette's "actual" human body—her woman form—escapes the logocentric contempt for embodiment, since this body is easily and fraudulently copied. Nothing essential adheres to a body, Thomas's ballet seems to tell us. This is even more strikingly clear in his interpretation of the climax, in which bodily death equals liberty. Therefore, it is safe to say that his version of *Swan Lake* is "visceral" only to the extent that it enacts an aversion of and an escape from viscera, a disdain for the mortality of the animal bodies of humans, which will become even clearer in Thomas's assessment of Nina after she begs for the part.

The actual audition is interesting for a number of reasons, not least because the viewer can see Nina dance for the first time.[14] She's dancing the White Swan part and the audience (which includes the viewer, Thomas, and other dancers) is supposed to be impressed by her performance, so that Thomas tells her that if he were only casting Odette, she would clearly be the right choice—but alas, subjectivity is not so simple, and Nina as the subject must struggle with her double. As is expected, the audition room is surrounded by mirrors, where we can constantly see Nina's reflections as the camera moves around her. Most interestingly, the camera does not show in the mirror: even though the cinematic point of view gets aligned with a mirror more than once, the camera fails to be captured by specularity (Figure 11.2).

Whereas in the three-part mirror scene I mentioned earlier, the camera could have been concealed by a mere tilt of the mirror frame, during the audition it is clear that it has been digitally erased. Such refusal to allow the cinematic apparatus to be visually captured by reflective surfaces resonates with by now classic discussions in film theory regarding the place of the gaze in filmic discourse. Laura Mulvey, initially, but also others have argued that the seeing perspective of the camera is constituted (psychoanalytically) as a masculine gaze that objectifies female bodies. If we include the mirror stage in this psychoanalytic discussion, one may conclude that the masculine seeing apparatus refuses to undergo the series of resemblances, (mis)identifications, and visual (self-)seduction that constitutes "normal" subjectivity. In other words, if Nina cannot be cast only as the White Swan—if she has to come to terms with the double in the mirror who is both her and not her—it seems that masculine identity (here clearly aligned with Thomas') can forgo that in order to reach its home in the Symbolic, safe from haunting, visually

Figure 11.2 The camera has been digitally erased from the image so that it is not visible behind Thomas's reflection *(Black Swan).*

constituted embodiment. This gender-biased specularity is relevant both for *Black Swan*'s filmic nature but also for its operatic background. Insofar as romantic ballet and opera were clearly intertwined, *Swan Lake* is "about the death of women," as Catherine Clément argues in her *Opera, or, The Undoing of Women*. Opera, for her, and here one can see the import to male invisibility in *Black Swan*, is about the "countless forms in which men want and want not to hear the woman's voice ... to know and not know what she knows about men's desires."[15] In this light, Thomas's focus on Odette/Odile to the expense of the Prince is articulated with the invisibility of the camera—and of the father figure in the three-part mirror. Men, Clément would say, want not to have their own sexuality narrated,[16] or captured in the never-ending reflections of truth and image in the mirror stage.

As described above, Nina fails the Black Swan portion of the audition due to Lily's interruption (or to Nina's displacement onto Lily of her own limitations, triggered by the repression of her inner Black Swan), so she decides to try and convince Thomas to change his mind the next day. This scene is crucial not only because she does succeed in getting the part, but because during it Thomas expounds on his metaphysics of dance. We learn that he has no criticisms of Nina's technique, but that he has never seen her "lose herself" while dancing, and without such abandon she cannot dance the Black Swan. Nina "just want[s] to be perfect," but for Thomas, "perfection is not just about control. It's also about letting go. Surprise yourself so you can surprise the audience. Transcendence. Very few have it in them" (*Black Swan*).

Thomas' theories of art and dance are, unsurprisingly, akin to his take on *Swan Lake*: bodily technique, for him, can only take the dancer so far. Technique, and the body itself, is only a tool the dancer-artist uses in order to express an artistic vision. The Black Swan may be carnal and seductive, but Nina is, ironically, *too embodied* to play her, since Thomas believes she would need to sever her intimate connection to her own body in order, precisely, to transcend it. At this point, the rhetoric of animality seems confused and scattered along Thomas' discourse and mine. For Thomas in fact insists that Nina *touch herself* in order to unleash the sensuality she ought to portray. This is one of the first evidences of the fact that the supplementary work of arche-animality articulates so many textual elements in *Black Swan* that animal rhetoric seems to occupy many, diverging positions.

One of these positions is the aspect I discussed above regarding the metaphysical dualism of Thomas' libretto, in which animal bodies constitute nothing but cages for human souls. This is reworked by his exposition of his beliefs about art—what I call his belief in *transcendanse*.[17] *Transcendanse* is precisely the ultimate instrumentalization of (and contempt for) the dancing body so as to make it convey ideal artistic content. True virtuosic achievement transcends the body and lies outside it, even though it must initially stem from within, since some ("very few") "have it *in* them." Thus, the body contains that which will transcend it and thus make true art, all the while failing to be acknowledged as such an origin.[18]

This neologism *transcendanse* is closely related to another one, by Lacan, which will be especially relevant for reading Nina's becoming-swan. In "Subversion of the Subject and the Dialectic of Desire in the Freudian Unconscious," he refers to what he calls the animal's *dancity* ("density" + "dance," or, in French, *dansité, densité + danse*). While discussing the deception of Speech and of the signifier, Lacan concedes that animals are capable of some level of pretence, "to be found in physical combat or sexual display" or hunting, and is "deployed in *imaginary capture*, and is integrated into the play of approach and rejection that constituted the original *dance*, in which these two vital situations find their rhythm, and in accord with which the partners ordered their movements."[19] This animal possibility of pretence, however, is still essentially different from that of a human subject, who is, according to Lacan, a subject of the signifier. That means that, while animals are *captured* in the density of the Imaginary's dance, human deception belongs to the Symbolic. This formulation is, of course, deeply interrelated to his conceptualization of the mirror stage as an apparatus of Imaginary capture.

Derrida reads Lacan as positing the never-ending specular capture on the side of animality: "The passage from imaginary to symbolic is determined as passage from the animal to the human order. [...] Whatever he concedes to the animal in this way [ability to feign, etc.], Lacan holds it in the imaginary or the presymbolic. [...] He holds that the animal holds itself in this captivity [of specularity]."[20] The passage into the symbolic means that the human is a subject of the signifier. But "the subject of the signifier is subject*ed* to the signifier. [...] The *real* human sovereign is the signifier. [...] This sovereignty is the superiority of man over beast, even if it is based on the privilege of the defect, lack, or fault, a failing that is referred to the generic prematuration of birth [neoteny] as well as to the castration complex."[21]The animal lacks the lack, and for this reason remains in the Imaginary realm, since the Symbolic is not needed to cover over a lack. This originary defect of the human, known in biology as neoteny, which throws him or her against the signifier, is read by Derrida "in the history of the original fault, an original sin that finds its mythical relay in the Oedipus story, and then its nonmythical relay in the 'castration complex' as formulated by Freud."[22]

This reference to an original sin connects easily to Derrida's discussion of the biblical Genesis in *The Animal That Therefore I Am*. The shamed nakedness before the cat which opens the book is articulated with the nudity revealed to Man in the biblical Fall. The human would then feel ashamed before the animal gaze precisely for lacking, for being incomplete and defective to the point that it needs the supplementary support of the signifier, technics, clothing, language, and so on. The human necessity of technical supplementation is described by Derrida as "a kind of *animalséance*: the single, incomparable and original experience of the impropriety that would come from appearing in truth naked, in front of the insistent gaze of the animal."[23] The incompleteness of humanity before both God and animals is felt, as Derrida points out, at the moment of original sin, the Fall, the realization of nudity. "Clothing derives from technics,"[24] he tells us, and Adam and Eve's covering of their nakedness reveals the need to cover a fault which they cannot perceive in the animals that surround them. Not only that, however, but this supplementation to their unfinished natural state sits side-by-side to Adam's naming of the animals, his experiment with language which sets him apart from them. As such, both language as revealed in the naming of the animals and clothing as revealed after the Fall emerge as two aspects of the same technical supplementation required by the animal who was not created as perfect as the other animals.

Black Swan concerns itself precisely with this interface between technics or technique and a human body which is both like and unlike animals' bodies. A. O. Scott, in *The New York Times* writes that "the subject of *Black Swan* [...] is the relationship, in art, between technique and emotion." However, the status of *emotion* here, in relation to the animality of the body, remains unclear: if opposed to the mechanicity of technique it can register as profoundly human; as opposed to humanist rationality, though, emotion promises an inhuman, carnal realm of possibilities. The latter formulation is explored in depth by Ariel Osterweis in her "Disciplining *Black Swan*, Animalizing Ambition," in which she places technique firmly on the side of the human world, and emotion and *ecstasy* on the side of carnality and animality. Therefore, she argues,

> *Black Swan* portrays artistic ambition through Nina's—and Odette/Odile's—erratic transformation from human to animal. [...] Aronofsky draws out the tension between technique's mechanical, disciplining function and the refined, animalistic characteristics of ecstasy. In doing so, he recognizes the idea that technique is often thought to provide formal tools with which an artist can express carnal emotion. [...] That Nina grows larger-than-life black feathered wings during the culminating *fouetté* turn sequence of *Swan Lake* [...] is significant, marking the point at which the pinnacle of her technical achievement coincides with raw animalistic attributes associated with the ecstatic. [...] Nina and her own "other" evoke the struggle of the artist to achieve balance between the human capability to master mechanical technique and the untamed animalistic emotion necessary to sustain ambition and evoke passion onstage.[25]

Osterweis seems to second Thomas in her belief that an artistic truth must be extracted from within Nina's technically mobilized body,[26] and that becoming-swan is the signifier of its liberation. Interiority, however, functioned according to the opposite logic in Thomas' account of *Swan Lake*, since there we had the story of a human being liberated from within an animal. Also, both Thomas and Osterweis seem to overlook the logic according to which the animalistic and the mechanical share the same space as the non-human, as that which operates automatically, unstoppably, and is instrumentalized by human intention. In short, they ignore both Descartes' conflation of animals and machines and the logic of supplementarity that grounds it: to human spiritual intentionality, both embodiment itself and machinery represent prosthetic supplements. The confusing status of some terms in the chain of supplements, such as *carnality*, *emotion,* and *ecstasy*, is explored in depth by Derrida in *Of Grammatology* with respect to Rousseau, but may be tracked in the Cartesian discussion at hand: both the mechanicity of technics and spontaneous ecstasy may figure either prior to or after humanity.[27] I believe Osterweis exposes this work of supplementarity when she refers to "bodily technique," crucially when comparing it to techniques of the cinematic apparatus:[28] if technique may be bodily, if human's "animal component" may be an instrument of technique, it is not clear how this same animal embodiment may give rise to carnality *as the other of* technics. This philosophical promiscuity, as it were, of the concept of animality in signifying practices (such as in *Black Swan*) is, I have argued, an effect of the repression of arche-animality, by means of which the vulgar concept of the animal is produced and textually scattered for diverse rhetorical uses.

A MULTIPLICITY OF FEATHERS

Returning to the psychoanalytic terms in which Deleuze and Guattari read becoming-animal in *Willard*, we could say that this discussion regarding becoming-swan goes to the heart of the role of animals in psychoanalysis, and of the extent to which they participate in the structure of the psyche and of signifying systems. Undoubtedly, Sigmund Freud's richest exploration of the signifying power of animals is also one of his most famous works, *From the History of an Infantile Neurosis*, also known simply as *The Wolf Man*, but spatial constraints mean I cannot fully discuss his theorization.

For the purpose of reading the psychoanalytic import of becoming-swan, one may foreground the intertextuality of *The Wolf Man*. Four "texts" are linked by Freud in an intertextual chain: the dream about wolves thanks to which his patient Pankejeff got his "Wolf Man" moniker, the fairy-tale "The Wolf and The Seven Little Goats," a children's book illustration depicting the wolf from this tale, and what Freud calls "the wolf story." For Freud, this latter story establishes the relation between tails and the phallus, with a tailless wolf standing in for a castrated wolf. In the dream, however, the wolves display peculiarly bushy tails—a form of compensation, argues Freud. But if bushy tails can stand in for taillessness that should mean that the distinction on which sexual difference is based—the main source of anxiety for the patient—is a false one. The different states of the tail should be the only stable signifiers, for without them that which supposedly matters in the primal scene the patient had supposedly witnessed—the reality of castration—loses all meaning. After deciphering the wolves thus, we can say that the wolves do not mean anything. All they can "mean" is the possibility of meaning, which is then Oedipally collapsed into the two signifying possibilities conceivable to psychoanalysis, castration and non-castration, similarly to the way *différance* is collapsed into the simple difference between writing and speech.

It is in what one may call Nina's becoming-animal drive (interpreted psychoanalytically as simply her inner Black Swan) that we can trace the similarities between *Black Swan* and *The Wolf Man*. If in *The Wolf Man* the signifying power of the wolves was collapsed into two possible meanings (castrated and non-castrated), the same is true regarding *Black Swan*. Initially, the duality in question might obviously appear to be between the White and the Black Swans, but I believe this is a misleading path. For one, one could say that this is the distinction at stake in *Swan Lake*, and that *Black Swan* is a different text. *Black Swan*, as a text, is much more invested, as I hope to have shown, in the contrast between *transcendanse* and *dancity*, where the former represents the disdain for and instrumentalization of embodiment in the pursuit of something "greater," and the latter the image of animality as an undifferentiated, dense materiality, where one is lost and undistinguished "as water in water."[29] As I hope it is now clear, these are the two most common answers to the problem of animal embodiment—*transcendanse* would be the standard Cartesian stance, whereas *dancity* would be the name for the materialistic approaches to animal. Both are the two sides of the same linguistic conceptuality and are secondary products of arche-animality. As it is, I have already shown how becoming-swan does in fact function to represent both signifying procedures.

However, we can return to a psychoanalytic interpretation—and confirm that Black/White is derivative of *transcendanse/dancity*—by focusing on the textual strata of the film. As in *The Wolf Man*, becoming-swan is *supposed* to be interpreted as a translation of Nina's psychosexual development as influenced by the *Swan Lake* libretto. Whereas the primal scene apparently inaugurates Pankejeff to the world of sexual difference as based on castration, Nina's psychosexual development similarly frames reality in terms of a duality. Girl/woman, mirror/reality, Imaginary/Symbolic are all related to the primordial psychoanalytic distinction between presence and absence of the phallus, castrated female and non-castrated male. Nina's mother's phallic-ness and cherishing of her daughter *as* a vicarious phallus work together to prevent Nina from entering the "correct" understanding that only males are phallic, that females are castrated, that the seduction of mirror images in the Imaginary is unimportant, and that only subjection to the signifier in the Symbolic constitutes subjectivity. These last two points reveal the complicity between the psychoanalytic material and the arche-animal themes of *transcendanse*—the overcoming of reflections, and their resulting invisibility, by means of entry into the Symbolic—and *dancity*—Imaginary capture in specularity.

In the film, the issue of castration as a hinge for Oedipus' textual coherence is presented in Nina's scratch on the back, a compulsively self-inflicted wound. This scratch is ambiguous, since it could, on the one hand, be read as the castration wound that normative femininity should learn to accept, meaning that Nina resists her mother's infantilizing influence by castrating herself, in order to finally actualize womanhood (i.e., the Black Swan). Ana España, in her "El Doble y el Espejo en *Cisne Negro*," suggests as much when she argues that Nina's scratching works as a form of relief from maternal influence, as well as an "onanistic substitute."[30] The crucial moment of auto-affection—as is the case of masturbation—has been productive ground for philosophical discussions, as can be attested by Derrida's reading of Husserlian auto-affection in the form of "the voice that keeps silence."[31] Nina's relationship to herself in masturbation, however, cannot help but go through the moment of specularity or otherness (the distinction here being irrelevant): this is revealed not only in the fact that the scratch is only visible to herself in a mirror's reflection,[32] but also in her explorations of sexuality, which are haunted by her doubles.[33]

On the other hand, the scratch can also be rightly read as the vehicle for becoming-swan which, as we shall see, shuns the distinction between castration and phallus. This means that castration registers not as the scratch wound, but as the attempts by both Nina and Erica at preventing further scratching. Nina's nails emerge as a phallic power that threatens Erica's influences, rendering their clipping, by both of them, as a sort of castration. Nina's nail-clipping is, in fact, interrupted by the "Black Swan" as it takes control while Nina holds the scissors in front of the bathroom mirror. Shown in the reflection, her face becomes that which the viewer comes to associate with the "evil" Swan, and she ends up cutting her own finger. España frames this as the Black Swan's attempt at "halting auto-castration—and thus allowing the growth of the scratch."[34]

It is in this light that it becomes crucial to inspect what consequences becoming-swan has for the reality of castration and for normative Oedipal sexuality. As it is, the scratching wound is the ideal signifier for the duality of castration, but the entire project of symbolizing such duality fails. The scratch plotline reaches its climax when Nina feels movement underneath it, sees (in the mirror) little black bristles pushing out, and finally fishes out a single black feather, foreshadowing her becoming-animal, but also explaining the presence of the scratch, as it is indeed irresistible to think its position signifies she will sprout wings out of her back. *One* black feather, *one* scratching wound: the pillars of Freudian poetics. However, becoming-swan multiplies penises and vaginas, and extends to Nina's whole body the reality of becoming: after this scene, corporal transformations include legs bent backwards like a bird's, elongated neck, and interdigital webs. Crucially, during her bravura performance of the Black Swan in the opening night, after having been completely taken over by becoming-swan, her entire skin ripples with bumps and valleys, all castrated wounds from which phallic feathers emerge. The multiplication of the unitary master concepts of psychoanalysis (phallus, vagina) is presented by Deleuze and Guattari as one of the main aspects of becoming-animal's multiplicity, which they connect to the Wolf Man both in regards to the wolf dream and his other symptoms.[35] They write:

> Salvador Dali, in attempting to reproduce his delusions, may go on at length about THE rhinoceros horn; he has not for all that left neurotic discourse behind. But when he starts comparing goosebumps to a field of tiny rhinoceros horns, we get the feeling that [...] we are now in the presence of madness. Is it still a question of comparison at all? It is, rather, a pure multiplicity that changes elements, or *becomes*. [...] The little bumps "become" horns, and the horns, little penises. No sooner does Freud discover the greatest art of the unconscious, this art of molecular multiplicities, than we find him tirelessly at work bringing back molar unities, reverting to his familiar themes of *the* father, *the* penis, *the* vagina, Castration with a capital C. [...] We're not far from wolves. For the Wolf-Man, in his second so-called psychotic episode, kept constant watch over the variations or changing path of the little holes or scars on the skin of his nose.[36]

And the Wolf Man dreamed about "six or seven" wolves, but, as we saw, they had "to be purged of their multiplicity."[37] Therefore, in Freud, "the little scars, the little holes, become subdivisions of the great scar or supreme hole named castration; the wolves become substitutes for a single Father."[38] I cannot stress enough that Nina's multiplicity of phalluses *is not merely an illustration* of a philosophical concept by Deleuze and Guattari. As I have argued, *Black Swan* as a text concerns itself entirely with the mapping and manipulation of Nina's body as the site of tension between different accounts of animality (how does the human "inhabit" his or her animal body, and what can this animal *mean*?) and different sexual realities (how real and necessary is castration?). These sexual realities, moreover, could indeed be termed *sextual* possibilities, since it is in the wider drama of the castration complex—including the Imaginary order, the mirror stage as the promise of bodily plenitude, and, finally its relinquishment in the Name of Father and the Symbolic for the master signifier, the phallus—that linguistic meaning as commonly understood takes shape. Thus, the diffraction of arche-animality or becoming-animal across Nina's body engages in a literality—actually rejecting Oedipal projects—that abandons the entire field of philosophical illustration.

As with *The Wolf Man*, a bidirectional line of influence can be drawn for *Black Swan*, where, besides the standard interpretation of swans as metaphors for sexual personae, the signifying power of the becoming-animal, indifferent to the distinction between phallus and castration, *transcendanse* and *dancity*, furnishes the meanings which are inscribed in the libretto of *Swan Lake* which, in turn, fuels the Oedipal drama. It should not be so shocking to suggest that a "fictional" text such as *Swan Lake* should influence psychosexual development, if only one remembers that the Oedipus complex is itself based on a play. Beyond that, this schema helps uncover a certain fictionality of the standard Freudian psychosexuality.

It is thus fitting that the climax of becoming-swan is precisely Nina's dancing of the thirty-two *fouettés* during the opening night. After giving full reign to her becoming, all the excitations of her skin—the field of vaginas and penises—gradually sprout black feathers with each turn, until her arms have transformed into vast black-feathered wings. This fuller transformation lays bare the textual aporias scattered in the text: is Nina really gradually becoming a bird? Have all the others dancers actually acquired her face temporarily? Will the audience flee in riotous horror at the monstrosity taking shape on the stage? More than anything, it is the reaction of the audience which will determine the "meaning" of becoming-swan, whether it will be collapsed into any of its possible significations or remain as pure potentiality of signification. In a sense, the reality of monstrous transformation—a horrified audience—would endorse *dancity*: the truth of animality is its body; meaningless material density constitutes reality; consciousness is an effect of matter. The other option, applause, confirms that Nina has been hallucinating all along, and sides with *transcendanse*: animals and their bodies (even the ones human inhabit) are only metaphors, instruments put to use by a psyche, either conscious or unconscious; there is no reality outside mental life.

The audience applauds. Thomas has won.

There is a remainder, however: she stands with bare arms, as a completely unfeathered human, but her shadows still reveal her wingspan. With this image the film offers a glimpse of what it could look like if becoming-swan were not collapsed into Oedipal alternatives. A much braver film would end before the audience provides its answer. All *Black Swan* does is show us the tragic consequences of this answer. All that follows the climax of becoming-swan rehearses its reduction into normative Freudian sexuality. Becoming-swan can now mean nothing but the simply sexual Black Swan, who can vanquish the virginal White one and make it accept female castration and the male phallus as the price of entry into Symbolic normalcy.

I have argued in this chapter that the truth of animality—*arche*-animality—cannot be said to be reducible to linguistic meaning. It produces no specific signification, despite still having meaning. This springs from the difference, highlighted by Barthes, between signification and what he calls *signifiance*: the former "belongs to the level of the product, of the statement, of communication."[39] *Signifiance* constitutes

> the signifying work, which belongs to the level of production, enunciation, symbolisation. [...] [It] closely resembles the dream-work, such as Freud began to describe it. [...] What 'dream-work" and 'text-work" have in common [...] is that they are a labour outside exchange, inaccessible to "calculation."[40]

Derrida, too, underscores the commonality between *signifiance* and Freudian dream-work. In "Freud and the Scene of Writing," he writes:

> The absence of an exhaustive and absolutely infallible code [for dream symbols] means that in psychic writing, which thus prefigures the meaning of writing in general, the difference between signifier and signified is never radical. Unconscious experience [...] does not borrow but produces its own signifiers; does not create them in their materiality, of course, but produces their status-as-meaningful (*signifiance*). Henceforth, they are not longer, properly speaking, signifiers.[41]

As both *Black Swan* demonstrates in its working, and I demonstrate in my analysis of its "animality," the animal can and indeed does break away from text. Not to affirm the primacy of non-textual matter, however, but to locate a site which comes in principle before this distinction, which makes it possible and that cannot, therefore, mean anything. I believe Derrida follows similar paleonymic procedures when he writes about the encounter with a real animal in *The Animal That Therefore I Am*. He assures us that the cat whose gaze he meets is a real, individual cat, and not a sort of ambassador for all cats or even all animals. It is precisely the unsubstitutable singularity of the animal who gazes that lends it an abyssal look. The existence of the animal one encounters thus "refuses to be conceptualized."[42]

However, this resistance from the cat against conceptualization must not mean a simple immanent materiality. Derrida's suspicion of simple materialism is inscribed in his whole oeuvre, and can be seen in his readings of Husserl, in the trace, in iterability, in arche-writing, and so on. In *The Animal That Therefore I Am*, he makes it clear that a more appropriate focus of discussion is the *animot*. Rather than either the material, singular, immanent animal (*l'animal*, lost within its own *dancity*), or the whole of the animal kingdom as an abstract concept (*les animaux*, before whose *transcendanse* each individual animal would be sublimated), *l'animot* articulates the singular and the plural, the material, and the ideal. It is "neither a species [...] nor an individual."[43] After all, the otherness of the gazing animal—however singular—must be registered *as* animal for the abyssal experience of the ends of the human to take place. This means that this encounter with the Other is an encounter with the *animot* as a hybrid of singular individual and generic species.

Thus, I argue that in *The Animal That Therefore I Am* Derrida relies on his previous work on paleonymy in order to conceptualize the *animot*, regardless of how subtly this reliance is conveyed. An awareness of this intertwining recognizes the extent to which the early Derrida had already addressed the "animal question" and, by the same token, the indebtedness of his late texts to his early deconstructive work.

NOTES

1. Giles Deleuze and Félix Guattarri, "1730: Becoming-Intense, Becoming-Animal, Becoming-Imperceptible...," in *A Thousand Plateaus: Capitalism & Schizophrenia* (London: Continuum, 1987), 233. Henceforth referred to as "Becoming-Animal."

2. *Black Swan*, directed by Darren Aronofsky (2010; Los Angeles: 20th Century Fox Home Entertainment, 2011), Blu-ray.

3. Rodolfo Piskorski, "Of Zoogrammatology as a Positive Literary Theory," *Journal of Literary Theory* 9, no. 2 (2015).

4. Leonard Lawlor, *This Is Not Sufficient: An Essay on Animality and Human Nature in Derrida* (New York: Columbia University Press, 2007), 30.

5. Ibid.

6. Ibid., 30–1.

7. Jacques Derrida, *Of Grammatology*, trans. Gayatri Chakravorty Spivak (Baltimore: John Hopkins University Press, 1976), 60, and Jacques Derrida, "Positions: Interview with Jean-Loius Houdebine and Guy Scarpetta," in *Positions*, trans. Alan Bass (London: The Athlone Press, 1987), 37–96, 64–5.

8. It seems relevant, too, that Nina's reflection appears much closer to her mother than the latter's to herself.

9. The execution of thirty-two consecutive *fouettés en tournant* is a traditional ballet move now strongly associated with virtuosity and technical achievement. It was originated by Italian dancer Pierina Legnani in 1893 in Marius Petipa's choreography for Fitinhoff-Schell's *Cinderella*, in St. Petersburg. After that, it was famously included in Petipa's choreography of *Swan Lake*, being therefore consecrated as the pinnacle of balletic technique from then on.

10. Jacques Lacan, "The Mirror Stage as Formative of the Function of the I as Revealed in Psychoanalytic Experience," in *Écrits: A Selection*, trans. Alan Sheridan (London: Routledge, 2001), 1, my emphases.

11. Ibid.

12. The wizard who, in the ballet, both turns Odette into a swan and who later fashions his own daughter into the former's likeness—creating the Black Swan—so as to trick the Prince.

13. Later, one of his directions to Nina performing the Black Swan is to "seduce us, not just the Prince, but the court, the audience, the entire world."

14. Natalie Portman's ballet training before filming received considerable media coverage, and may have been responsible for her Academy Award for Best Actress. However, it was probably known to most viewers that special effects and camera

tricks were used to create the impression she could actually dance ballet, while she in fact used a dance double, Sarah Lane, in the most technically challenging scenes. Ironically, for a film concerning the danger posed by understudies, a controversy originated regarding the extent to which Lane's work was not given full credit: an entire Wikipedia page is dedicated to this polemics "Black Swan dance double controversy," *Wikipedia, The Free Encyclopedia*, accessed 1 May, 2016. https://en.wikipedia.org/wiki/Black_Swan_dance_double_controversy). The double Lane claims she was even asked by producers not to give interviews before the Oscars ceremony, presumably to preserve the fantasy that Portman had indeed danced all her scenes (Christopher John Farley and Sarah Lane, "Natalie Portman's 'Black Swan' Dance Double Says She Deserves More Credit," *The Wall Street Journal*, last modified 26 March, 2011, http://blogs.wsj.com/speakeasy/2011/03/26/natalie-portman's-black-swan-dance-double-says-she-deserves-more-credit).

15. Catherine Clément, quoted in Cary Wolfe, "When You Can't Believe Your Eyes (or Voice): *Dancer in the Dark*," in *What is Posthumanism?* (Minneapolis: University of Minnesota, 2010), 174.

16. Matthew Bourne's version of *Swan Lake*, with a Prince protagonist and an all-male cast of swans, would seem to offer the precise opposite possibility: the thematization of male sexuality. Its queer aesthetics, therefore, seems to stem from the very fact that masculinity is *not* made invisible—visibility thus both objectifies and eroticises.

17. Thomas Leroy is a French character—I believe that he would hear the word *danse* in the French word *transcendance*.

18. In his analysis of the film, Pietro Bianchi frames *Black Swan* as a discourse on the body as that which is the *object* of dance as an art form, instead of its subject. For him, "dance thematises [...] the division that traverses any subject: if the body is the *object* of dance, who is its subject?" (Pietro Bianchi, "La Ferita del Desiderio," *Cineforum* 502 (2001), 32, my translation). To all purposes, the dance subject is outside it, the director or choreographer, "an Other, [...] who would like to reduce the body to an objectified element of representation, played in this film by Vincent Cassell," (ibid., my translation).

19. Jacques Lacan, "The Subversion of the Subject and the Dialectic of Desire in the Freudian Unconscious," in *Écrits: A Selection*, trans. by Alan Sheridan (New York: Norton, 1977), 305.

20. Jacques Derrida, *The Beast and the Sovereign, Volume 1*, trans. by Geoffrey Bennington (Chicago: University of Chicago Press, 2009), 125, 122.

21. Ibid., 125.

22. Ibid., 134.

23. Jacques Derrida, *The Animal That Therefore I Am*, ed. Marie-Louise Mallet, trans. David Wills (New York: Fordham University Press, 2008), 4.

24. Ibid., 5.

25. Ariel Osterweis, "Disciplining *Black Swan*, Animalizing Ambition," in *The Oxford Handbook of Dance and the Popular Screen*, ed. Melissa Blanco Borelli (Oxford: Oxford University Press, 2014), 71, 74.

26. She writes: "Human body parts are mangled and removed, and what remains are wounds *from which* the animalistic *emerges*" (ibid., 73, my emphasis).

27. Technicity may be seen as present in the automatic world of animal instinct, which the advent of humanity breaks by means of reason and intentionality, but

may also be feared/adored as that which comes to supplement, via culture, an imperfect human nature. On the other hand, carnality can be located in the immanence of animal-being, consequently interrupted by human rationality, or as a posterior remedy for corrupt, overly mechanised humanity, as Osterweis suggests is Nina's case.

28. "In terms of virtuosity, *Black Swan* creates a parallel between the cinematic apparatus (the full schematic technicity of cinema, including camerawork) and the dancer's bodily technique, an alternating concealment and exposure of the mechanical" (Osterweis, "Disciplining *Black Swan*," 74).

29. Georges Bataille, "Animality," in *Theory of Religion*, trans. Robert Hurley (New York: Zone Books, 1989), 19.

30. Ana España, "El Doble y el Espejo en *Cisne Negro*," *Fotocinema: Revista Científica de Cine y Fotografía* 4 (2012), 130, my translation.

31. Jacques Derrida, "Speech and Phenomena: Introduction to the Problem of Signs in Husserl's Phenomenology," in *Speech and Phenomena and Other Essays on Husserl's Theory of Signs*, ed. Newton Garver, trans. David B. Allison (Evanston: Northwestern University Press, 1973), 3–106.

32. Ibid., 131.

33. Her attempt at "touching herself" as instructed by Thomas happens in the presence of her (hallucinated?) sleeping mother, and her sexual experience with Lily is framed by the film as fantasy or, in other words, as auto-eroticism channelled through Otherness.

34. España, "El Doble," 132, my translation.

35. "The Wolf-Man's pack of wolves also becomes a swarm of bees, and a field of anuses, and a collection of small holes and tiny ulcerations," (Deleuze and Guattari, "Becoming-Animal," 249).

36. Giles Deleuze and Félix Guattari, "1914: One or Several Wolves?" in *A Thousand Plateaus: Capitalism & Schizophrenia* (London and New York: Continuum, 1987), 27–8, emphases in the original.

37. Ibid., 28.

38. Ibid., 31.

39. Roland Barthes, "Theory of the Text," trans. Ian McLeod, in *Untying the Text: A Post-Structuralist Reader*, ed. Robert Young (London: Routledge & Kegan Paul, 1981), 37.

40. Ibid., 37–8, 40.

41. Jacques Derrida, "Freud and the Scene of Writing," in *Writing and Difference*, trans. Alan Bass (Chicago: The University of Chicago Press, 1978), 209. Interpolation of the original French is the translator's.

42. Derrida, *The Animal That Therefore I Am*, 9.

43. Ibid., 41.

WORKS CITED

Barthes, Roland. "Theory of the Text." Translated by Ian McLeod. In *Untying the Text: A Post-Structuralist Reader*. Edited by Robert Young, 31–47. London: Routledge & Kegan Paul, 1981.

Bataille, Georges. "Animality." In *Theory of Religion*. Translated by Robert Hurley, 17–26. New York, NY: Zone Books, 1989.

Bianchi, Pietro. "La Ferita del Desiderio."*Cineforum* 502 (2001): 32–34.

Black Swan. Directed by Darren Aronofsky. 2010. Los Angeles, CA: 20th Century Fox Home Entertainment, 2011. Blu-ray.

BDeleuze, Giles, and Félix Guattarri. "1730: Becoming-Intense, Becoming-Animal, Becoming-Imperceptible. . . ." In *A Thousand Plateaus: Capitalism & Schizophrenia*, 233–209. London: Continuum, 1987.

———. "1914: One or Several Wolves?" In *A Thousand Plateaus: Capitalism & Schizophrenia*, 26–38. London: Continuum, 1987.

Derrida, Jacques. *The Beast and the Sovereign, Volume 1*. Translated by Geoffrey Bennington. Chicago, IL: University of Chicago Press, 2009.

———. "Freud and the Scene of Writing." In *Writing and Difference*. Translated by Alan Bass, 196–231. Chicago, IL: The University of Chicago Press, 1978.

———. "Speech and Phenomena: Introduction to the Problem of Signs in Husserl's Phenomenology." In *Speech and Phenomena and Other Essays on Husserl's Theory of Signs*, ed. Newton Garver, trans. David B. Allison. Evanston, IL: Northwestern University Press, 1973.

———. *The Animal That Therefore I Am*, ed. Marie-Louise Mallet, trans. David Wills. New York, NY: Fordham University Press, 2008.

———. *Of Grammatology*, trans. Gayatri Chakravorty Spivak. Baltimore, MD: John Hopkins University Press, 1976.

España, Ana. "El Doble y el Espejo en *Cisne Negro*."*Fotocinema: Revista Científica de Cine y Fotografía* 4 (2012): 120–139.

Farley, Christopher John, and Sarah Lane. "Natalie Portman's 'Black Swan' Dance Double Says She Deserves More Credit."*The Wall Street Journal*. Last modified March 26, 2011. http://blogs.wsj.com/speakeasy/2011/03/26/natalie-portman's-black-swan-dance-double-says-she-deserves-more-credit.

Lacan, Jacques. "The Mirror Stage as Formative of the Function of the I as Revealed in Psychoanalytic Experience." In *Écrits: A Selection*. Translated by Alan Sheridan, 1–6. London: Routledge, 2001.

———. "The Subversion of the Subject and the Dialectic of Desire in the Freudian Unconscious." In *Écrits: A Selection*. Translated by Alan Sheridan, 292–325. New York, NY: Norton, 1977.

Lawlor, Leonard. *This Is Not Sufficient: An Essay on Animality and Human Nature in Derrida*. New York, NY: Columbia University Press, 2007.

Osterweis, Ariel. "Disciplining *Black Swan*, Animalizing Ambition." In *The Oxford Handbook of Dance and the Popular Screen*. Edited by Melissa Blanco Borelli, 68–52. Oxford: Oxford University Press, 2014.

Piskorski, Rodolfo. "Of Zoogrammatology as a Positive Literary Theory."*Journal of Literary Theory* 9, no. 2 (2015): 230–249.

Wolfe, Cary. "When You Can't Believe Your Eyes or Voice: 'Dancer in the Dark'." In *What is Posthumanism?* 169–202. Minneapolis, MN: University of Minnesota, 2010.

Chapter 12

Approaching Apocalypse

The Typology of Animals in Nicola Barker's In the Approaches

James Tink

One of the most intriguing moments in Derrida's exercise in his animal autobiography is when his celebrated encounter with the pet cat points him to the end of the world. The uncanny experience of being observed by the animal other provokes a consideration of the limits of the human, which leads to a thought of the apocalypse.

> As with every bottomless gaze, as with the eyes of the other, the gaze called "animal" offers to my sight the abyssal limit of the human: the inhuman or the ahuman, the ends of man, that is to say, the border crossing from which vantage man dares to announce himself to himself, thereby calling himself by the name that he believes he gives himself. And in these moments of nakedness, as regards the animal, everything can happen to me, I am like a child ready for the apocalypse, *I am (following) the apocalypse itself*, that is to say, the ultimate and first event of the end, the unveiling and the verdict. I am (following) it, the apocalypse. I identify with it by running behind it, and after it, after its whole zoo-logy. When the instant of extreme passion passes, and I find peace again, then I can speak calmly of the beasts of the apocalypse, visit them in the museum, see them in a painting … I can visit them in a zoo, read about them in the Bible, or speak about them as in a book.[1]

Derrida's prose (even in this translation of the punning of the verb form *suis*) performatively evokes the uncanny phenomenon it describes: a glimpse of the gaze of the animal other is a momentary encounter with the wholly other as if it were an apocalyptic mode of thought, an unveiling of the animal as something other that is also the premonition of some end of the human, which is soon curtailed by the reassuring calculations of a familiar order of the same. This could also be said to describe an experience of the event as a brief sense of the impossible: of *l'arrivant* or the opening to an incalculable future that is returned to calculability and sameness. The main argument of the short *Animal* volume, and the later seminars on *The Beast and the Sovereign*, is of course how the incalculable event of the other—and especially all otherness associated with the animal, or more precisely, in Derrida's neologism, the philosophical *animot*—might help us redraw traditional assumptions of mastery and sovereignty that have been announced on behalf of the auto-affective (masculine) human. The apocalypse passage above, moreover, allows us to think of the animal in another mode of critical thought, which is that of typology.

Typology can refer to the general principle of classification (as in zoology), but more specifically to the Biblical hermeneutic of reading any figure or thing in Old Testament Scripture as signifying an additional meaning in accordance with the New Testament, and thence the revelation of the entire Christian salvational narrative, such as the crossing of the River Jordan signifying the later idea of baptism. As explored by Erich Auerbach, typology (or figural interpretation, to give its Latinate form) became the reading practice of interpreting events or things as signs prefiguring a later, fulfilled understanding of a sign or a narrative, whether in Christian belief or more secularized thought.[2] Derrida evokes both senses of typology in the passage, although not as a confirmation of Judeo-Christian thought. Derrida's later work, as is well-known, explores ideas of the trace, event, and future in terms of a weak thought of the "messianic without messianism";[3] the animal seminar, in addressing Emmanuel Levinas' concept of the Other as implicitly humanist, proposes the provocative idea of the "divinanimality" of the Other, as will be discussed below.[4] Nevertheless, Derrida's suggestive use of typology invites a way of seeing animals, particularly literary animals of fiction. One reason is, as Greg Garrard has argued, that the cultural repertoire of animal meanings and symbolism—be that anthropomorphic, zoomorphic, or otherwise—is itself available as an overall typology of figures and tropes, and that this provides the best way to catalogue the uses of animals in literature and culture.[5] Less directly, there is another literary use of typology to suggest ways in which narratives impose coherence and meaning to specific events and figures in a plot through the use of peripeteia and closure, and so follow the apocalypse by imposing revelation as the telos of narratives. As Frank Kermode has argued, concepts of past, present, and future provided by fiction may in fact rely on hermeneutic modes of revelation and typology, a "sense of the ending," that amount to secularized Christian concepts: "The theologians revive typology and are followed by the literary critics."[6] Derrida also invites literary critics to think again about narrative and the senses of the future, and this can be developed through his study of animals and *animot.* In this chapter, I am going to explore this possibility through a reading of a contemporary English novel that, felicitously enough, is concerned with both the problem of understanding the animal other and of understanding a typology, Christian or otherwise. This is Nicola Barker's *In the Approaches*, published in 2014 and longlisted for the UK Man Booker Prize. By doing so, I hope to explore the recent suggestion of Nicholas Royle that after Derrida, "The novel would be a singular and new kind of 'literary animal.'"[7] In fact, Barker's novel is, I will argue, a singular fiction about animals, and therefore I am going to give an analysis of these animal traces which refers back to Derrida's speculations.

Nicola Barker belongs to the generation of British novelists, championed by *Granta* magazine in their list of 2003, who emerged at the turn of the century, including Ali and Zadie Smith or the Davids, Mitchell and Peace.[8] Like Ali Smith, Barker's use of colloquial British speech, naïve character narrators, and occasional humorous whimsy seems a deliberate attempt to make the tone of the works more playful and exuberant than the crystalline, careful prose of other recent literary fiction such as John Banville.[9] *In the Approaches*, her tenth novel, could best be described as a metaphysical romantic comedy set in 1980s small-town Southern England, which combines a tale of two squabbling, middle-aged misfits reluctantly admitting their attraction to each other, a quest to solve an old local mystery, and a more ambiguous account of the mystical visions and revelations of a child saint, which perhaps makes the novel a form of apocalyptic, divine comedy. Although 'The Approaches' actually refers to the coast of East Sussex, it also implies the idea of a journey, quest, and arrival that describes the movement from animosity to reconciliation in the plot. Summarizing a novel like this is not easy, but will be necessary. In the Orwellian year 1984, an American journalist, Franklin Huff, vacationing in the South Coast of England at the village of Pett Level, has a series of fractious and embarrassing encounters with local resident Carla Hahn. In fact, the cynical Huff is secretly investigating a local scandal from 1971, when a 12-year-old disabled Irish-Australian girl, Orla Cleary, who was acclaimed by some as a Roman Catholic saint with the power of visions, died in mysterious circumstances. Carla, who once used to nanny for the child, begins to recover long-repressed memories of 1971 (smells, the words of a prayer, the mysterious number 4004) and believes that the ghost of Orla is making herself present among the community again. While Franklin and Carla reluctantly squabble and draw together like some latter-day Benedick and Beatrice, Franklin recovers Orla's long-lost ancestral heirloom from a priest, an old Aboriginal coat which on inspection by a local scientist-turned-poet contains a complex mathematical equation relevant to computer science and what readers will recognize as the Internet. At the end of the novel, Franklin, by now fallen in love with Carla, apparently begins to experience a vision of light emerging from Carla's body and a prophetic glimpse of Orla's spirit regenerating the whole world. Barker's novel is thus an apocalyptic fiction, not in the sense of the destruction of the world, but because it describes an experience of revelation or unveiling of the future through the gradual approach of Orla to the cast of characters. It also poses a challenge for secular liberal readers by combining a potentially pious Roman Catholic symbolism of sainthood with a techno-scientific motif about the beginnings of the Internet and the digital revolution in the 1980s. The shifts in tone between ribald, comic realism and mystical piety also seem to be a deliberate test of the reader's own credulity toward the 1980s setting and their suspension of disbelief regarding the increasingly strange storyworld of the novel: a test of faith that is surely meant to be analogous to the characters' own readiness to ditch their skepticism in favor of an inner calling, be that of love or belief in spirits.

Yet crucially, the novel is also full of animals, which circulate in the text as agents, objects, and, most strikingly, a narrator. The novel is narrated by four alternating character-narrators, primarily Franklin and Carla, but also their neighbor Clifford, who operates as a meta-fictional presence in the text by commenting on the motives of Nicola Barker as the author of the work, and finally Teobaldo (Baldo) an African Grey parrot who has its own subplot as the unhappy pet of a middle-aged woman. A substantial part of the novel, then, is a recreation of the sense world of the animal. However, this is not the only trace of the literary animal: the text is in fact full of events where characters interact and respond to animals: Franklin and Clara's initial animosity stems from his accidentally injuring her cat and her leaving him with a rotten shark carcass (!) in revenge; Orla's symbolic coat is discovered by a dog. Animals are also prone to be read by characters as well as readers, as potential signs or symbols of an overall typology. Early on in the novel, Franklin observes a rescued rabbit: "I am no fan of small mammals. I have given it temporary berth in the bath. In the bath the enamel turns its white fur a yellower hue. Strange how the act of comparison can suddenly transform one clearly defined object into something else altogether. Life has a nasty habit of doing that."[10] As the novel progresses, animals play a surprisingly significant role as figures of transformation, and indeed of life. In what follows, we will consider some of the main ways in which the text imagines animals and other living beings.

One desire of animal-based literary studies is to extend the imaginative realm of fiction to include different situated and embodied types of knowledge and sensory *umwelt* in a shared storyworld.[11] In describing the sensations of Baldo, the female pet of Carla's best friend, "The eternally fragrant, sweet-natured and well-meaning Alys Jane Drury" (27), the novel adopts the focalization of the bird to represent its consciousness and view of the human world. Baldo is, *pace* Heidegger, not so much poor in the world as pissed off with it, as the British colloquialism would have it: ruing her captivity as a pet and an owner who mistakes her sex for male. These sequences are also license for the novel to play with typography and sentence structure, as we can see from just one lengthy quotation:

> *Scratch, scratch, scratch.* Feather, feather, feather! *Look!* Soft feather down like grey snow! Good! Good for nest. Oh. No.
> No nest.
> Poor Baldo
> *Hmmn.*
> 'Room'.
> 'Cage'. 'Chair'. 'Lamp'. 'Dresser'. 'Ceiling'. No sky! 'Celing'. No sky! Dead Sky. Gone sky. Can't … can't … ! No sky!
> Wanna fly.
> Sad moment.
> Whassat? *'Sun'*! Baldo, look! See 'sun'!
> Getting closer!
> Joy!
> *Baldo! Baldo! Baldo! Baldo!*
> Hmmmn.
> Egg.
> Why no egg?
> Why no nest?
> *Bounce! Bounce! Bounce! Bounce!*
> Baldo! Baldo! Baldo! Baldo! Wah! *Wah*! WAH!
> Oh … Uh-oh … Here she comes, here she comes. Jailer! Bitch! Here she comes!
> Bow, deep bow. Respectful. Deep bow. Baldo, Baldo, Baldo, Baldo …
> Away she goes again! Gone. Gone. Lonely Baldo. Ruffle feathers. Where's the …?
> 'Mirror'! Ring the 'bell'! Look in 'mirror'!
> WAH!
> Look! Look! Whosat? Whosat? Spirit parrot! Whosat? Eye! Evil! Beak! Sharp! Dead parrot! Ghost parrot! Whosat?
> *WAH!*
> *Ruuuun!* (34)

While this sequence is certainly focalized from the perspective of Baldo, it is not entirely clear whether this is supposed to be read as Baldo's direct speech or some form of free indirect speech, or why certain nouns are quoted as if a sign of acquired vocabulary as opposed to the "natural" speech of the animal. The bird may not understand the mirror image, but it does understand deception. The subplot involves the unwelcome arrival of a new unfriendly male mynah bird, King, whom Alys is pet sitting, and with whom Baldo cannot apparently communicate. Baldo plots to regain her dominion by the cunning plan of imitating the sound of a telephone deliberately badly, so as to trick the showoff King into mimicking it properly, which of course infuriates the harassed "Bitch" Alys enough to get rid of the upstart newcomer. Baldo celebrates this with sovereign laughter ("WAAAAAA …" (264)).

Nicholas Royle suggests, following Derrida, that the novel needs to recover a poetic dimension in order to rethink subjectivity as part of animality, and, as shown above, the novel certainly experiments with prose style to suggest the sensory world of the capricious, unhappy animal as a form of trickster figure.[12] Perhaps for some readers it might demonstrate the theme of the abuse of the neglected, captive animal; however, for others it might seem too cartoonish and childish a representation of the animal to be original, even though, in actual fact, it seems more like an enjoyably anarchic Warner Brothers style cartoon than the "disnification" criticized by some theorists.[13] Then again, Baldo's bad temper is in accordance with the other human characters such as Franklin, all of whom show an agitated, disgruntled, and comically frustrated outlook at the start of the novel. Moreover, the novel rather preempts charges of anthropomorphism through the metafictional intervention of Clifford, who criticizes the sequence:

> Although ... Although [...] what about that poor parrot? Baldie? Baldo? How's he/she fit into this mess? What did that blessed parrot ever do to anybody? Doesn't seem right—fair—to have his/her/our innermost thoughts—our private feelings and ideas (uninspiring as they most certainly are)—casually picked over (exploited, let's make no bones about it) for the sake of a little light relief. (85)

Is it simply light relief? And in any case, is the attempt to render the inner life of one animal character any less reasonable than the nominally human agents in the novel? At least the novel acknowledges what Timothy Clark has called the challenge of anthropomorphism to represent animal lives and senses without illusion of injustice.[14] Yet the role of the animal character acquires another significance in the novel through an additional, typological value as a sign portending the future. This happens in the strange sequence where Clifford describes a mystical vision on the beach (its precise epistemological status remains unclear) in which the animal is seen again from a different perspective. Clifford's narrative, broken down into fragments, describes characters mixing with angels is a state of harmony:

> [T]here were other people there who I ... who had been a part of ... like Alys Jane ... like ... like ... holding a little bird cage which was a ... a little screen ... of ... of feelings ... and there was a parrot ... a ... a *bird* ... inside it who when you pressed something softly-*lee-lee* he began to sing ... to caw ... to tweet-*eat-eat* and just ... he took to the wing and then he ... I can't really. (395)

What is the agency behind the "little screen of feelings": is it simply Aly's exercise of pity to the animal (and if so, is that a narcissistic and anthropomorphic use of the animal as a mirror) or could it be another encounter between the two singular entities themselves? Perhaps the song and flight of the bird seems to suggest a symbol of transcendence, recalling the golden bird that sings in W. B. Yeats' "Byzantium" for example. In this transformation scene, what matters most is that the visualization of the animal that involves a new, apocalyptic revelation of the animal as part of the fulfillment of the future state: this is the signature typological mode of the novel.

One way to understand this obscure moment is to compare it with some common literary representations of the parrot; in Derrida's terms, a thinking of the particular *animot* with which this creature has been used in literature. In his discussion of Defoe's *Robinson Crusoe,* Derrida considers the familiar fact of the bird's mimicry of human speech as being essentially a prosthetic device in which a fantasy of a sovereign human ipseity and *autos* could be engineered through the repetition of a voice to itself: Crusoe is reassured to hear his name recited back to him by Pol the pet parrot.[15] Perhaps the doting, but ignorant, relationship of Alys to Baldo could be read as a similar confirmation of the human agent to itself, except that there seems to be another, better intertextual reference available. This would be the relationship of good-natured Félicité to her parrot Loulou in Gustave Flaubert's short story "A Simple Heart" ("Un Coeur Simple," 1877), in which the simple devotion of the woman to the bird in the cage might strike sophisticated readers as sentimental and rather *bêtise*, until Félicité is rewarded with a vision of the bird leading her to heaven. Flaubert's story is a clever engagement with the expectations of realism, as well as a story about faith and unconditional ethical relationship to the animal, and something of the same effect is implied by Barker.[16] Moreover, in both stories the bird can also be best understood, not as a prosthesis, but as a version of what Derrida calls the potential "divinanimality" of the Other. Derrida suggests that in any symbolic order, any quasi-transcendental concept of the Other must necessarily be sufficiently nonhuman, and so potentially quasi-animal, to structure that very order. "This is one of the reasons why it is so difficult," he suggests, "to utter a discourse of mastery or of transcendence with regard to the animal and simultaneously to claim to do it in the name of God, in the name of the Father, or in the name of the Law. Must one not recognize Father, Law, Animal, etc. as being at bottom the same thing?"[17] Derrida's neologism, arriving late in the Animal seminar, seems deliberately provocative ("Is Derrida deifying cats?" Steven Shakespeare asks rhetorically[18]), and we must return to the implications of this theme later, but it provides a way to approach the strangeness of the animals in Barker's work, and its own peculiar sense of the apocalyptic.

By the end of the novel, pet-hating Franklin has been humiliated, physically injured on the buttocks (an affliction with no dignity, and thence suitably comic) and become obsessed both with Clara and the story of Orla. Like Cliff, he also experiences his own apocalyptic vision of the earthly paradise, with the saintly child Orla returning to a transformed Sussex:

> The child had briefly withdrawn, but then that soft sound returned: the broken … antique … broken bamboo chimes … although that hardly describes it … and I saw the dust in the furthest reaches of the room, settling on the heads of literally dozens of people … a dense sea of people … the dust making their presence manifest … all those ghostly shapes … holding out their hands to the dust, that they might be made … whole … real … for even the briefest of interludes. A terrible groan of conjoined yearning and love and pain and desire echoed around the room. I even saw a *dog*—a dog, a small shadow terrier, jumping up on its hind legs. It was with a woman who reached out to restrain it, whispering a muttered expletive … The entire room was … like it was exposed by a golden searchlight … and all these … should I call them spirits … souls? Reaching out their arms with this mournful sigh … then the laugh of the child (441)

It is again difficult for the reader to fully comprehend the passage. It could be Franklin's subjective hallucination or some prediction of a future event in the storyworld of the novel (if so, presumably not the year 1984 most of us remember). Perhaps it is paradise, but why is he incredulous about the presence of a dog? Carla interacts with pet dogs throughout the narrative, so it could be a displaced reference to her, the woman with whom he has fallen in love. In any case, it is as if the banality of the scene is supposed to be at odds with its significance, or that Franklin deems the presence of the bestial dog as somehow to be at odds with holy symbolism. As with the sight of the parrot, the animal takes on the importance of a sign or symbol prefiguring the future. Furthermore, Carla, also reads animals as traces of some hidden narrative design, including the innocuous rabbit rescued by Franklin:

> I suddenly realized that it had all happened—all of it, every, little bit of it, even the *rabbit*—for a reason. Everything was interconnected. We were all chosen. We. *Us*. You know—the dregs. The pointless. The flawed. The unconvinced. … We were all chosen. The least worthy in many respects. […] We were all chosen. Because majesty shines more brightly through … through frailty, I suppose (385; italics in original).

As a result of these details, the reader also begins to reread animal references as foreshadowing the plot outcome. For example, the fact that a dog discovered Orla's long-lost Australian Aboriginal coat among some rubbish is subsequently reconsidered as a profound event: "It was the coat! Orla's special coat! And the damn dog (or the blessed dog—the jury's still out) had this most precious of artifacts, this most … how can I explain it? The most rich and sacred and intimate and … and *rare* and talismanic of jewels caught between its jaws, and it was dragging it through the dust … . A dog … the dog had" (413). The series of animal references seems to pass beyond mere coincidence and begins to strike the reader as part of some underlying design.

If *In the Approaches* imagines a fictional world where the animals are traces of some imminent goal, then they are also used to evoke some other properties of knowledge. As mentioned above, one device of the plot is that Orla's relic—her coat—contains a mysterious code that is deciphered by one character, Sage Meadows, as a breakthrough equation for 1980s computer science. This is signposted, perhaps a bit too neatly, through the mysterious number 4004, which, as Meadows explains, refers to Archbishop Ussher's notorious estimate for the date of creation, but also the Intel 4004 computer chip, which inaugurated (or should that be ushered in?) the IT revolution: "'And I suppose with the Intel 4004 another kind of world was born,' Mrs Meadows expanded, 'a new world. A digital world'" (390). Yet crucially the moment of inspiration that triggers her realization is in fact seeing another animal, a swarm of bees, as she explains to Carla and Franklin:

> 'Did you see the bee?' Miss Hahn asked, pointing.
> 'Good heavens, yes! A bee! And the swarm is ... Oh my God! That's a stable Dyson swarm! A swarm! Next to the bee!'
> She turned to Miss Hahn, astonished, 'After we were talking this morning, Miss Hahn, and I was thinking about the hive, and I was ... something was coming together in my mind but I couldn't quite ... a kind of ... of *inspiration*. But there was something missing, something ... *undecidable*. It was *this*. Good heavens! I have this extraordinary feeling! In my stomach! A fluttering. A sense ... of *rightness*! Do you understand me, Miss Hahn? Mr ... Mr Huff? (428. Italics in original)

The reference is obscure, but while a 'Dyson swarm" refers to the idea from Frederick Dyson of an energy-saving spherical construct, it also evokes the reading of the insects as mathematical signs and shapes: Sage also describes the bees' movement as a "Hopf Bundle" and alludes to Internet research (430). The nonhuman element of this particular encounter with the other is a code as much as an animal, which invites further questions about the precise relation of human, nonhuman, and animality. At least in this particular strand of the novel, the apocalyptic effect of the animal is toward ideas of shape, number, and design. We need to finally reflect on what these "undecidable" experiences of the animal may mean.

"There is a certain *Triumph of Life* in *bêtise*," Derrida claims enigmatically in *The Beast and the Sovereign* (I 183).[19] That seminar, which continues the study of animality, also contains a thinking of comedy, or of the absurdity or *bêtise* that can arise when the human assumes sovereignty over the condition of animality. The novel could also be read as a comic triumph of life and sympathetic glance at folly: the rural English setting of The Approaches becomes a form of green world in which both humans and animals are restored to happiness.[20] This principle of regeneration is addressed in the novel at certain points as a persistence of life, even as a principle of animal life, as with Carla's observation:

> Because there's still life on the heap. Not ... not anything to get too ... you know, worked-up about, but still ... yes...still life. A second-rate kind of life. Life on a shoestring. Cut-price life. No frills. And tiny. Microscopic. Infinitesimal. Virtually invisible to the naked eye.
> But there's still ... still life. Of a sort. And heat (a tiny glow). And hope. (341)

Franklin also identifies this power of living on, much more bluntly: "'Indestructible!' I muttered. 'She's indestructible! Like a cockroach!'" (381). The metaphors of an insect or microscopic form of life suggest that a theme of the novel is the endurance and vindication of neglected life (here presented as a bare form of animality or prehuman life) at some point in the future, be that mystical transformation or more personal changes of circumstance. This also suggests whether Barker's novel could be read as part of a broader recent cultural turn towards idea of life and presence. Clare Colebrook, for one, has argued that the new century has been distinguished by a general intellectual turn from problems of the linguistic or post-structural to "affirmative turns to life, affect, bodies and the vital" in an overall "affirmation of life."[21] Could this turn to a notion of vital life be the best way to understand the novel?

Certainly the theme of enduring life points to the Utopian potential within the novel, both in its visionary, apocalyptic mode of a revealed paradise, and within the generic form of the green world comedy noted above. The tenor of the novel could best be called a *redemptive* one, in which a condition of life, shared between human and animals, is revealed to Clifford and Franklin as an earthly paradise. This also has some resonance with contemporary cultural theory. Giorgio Agamben's influential *Homo sacer* project includes some reflections on the human/animal distinction in *The Open*. That volume sketches a genealogy of modernity in which the concept of animality is said to have been severed from that of humanity, yet Agamben postulates a latent "Shabat of both animal and man" in which forms of life can be reconciled and reimagined.[22] It would be tempting to read Barker's novel as providing a fictional exploration of a similar concept of Sabbath, as in the blissful image of Alys and the bird together. Both Barker and Agamben also pursue an openly Judeo-Christian apocalyptic line of thought to imagine a vision of redeemed life. In these ways, Barker's novel could be called a sort of divine comedy, and also exemplifying a certain affirmation of life in contemporary thought.

However, thinking with Derrida would require a different approach to the novel and the idea of animality, one not so confident about identifying a continuum of life, or bare life, in the animal. This is because Derrida's main fascination with the animal throughout the late works is the way it problematizes assumptions about life and sovereignty and cautions against the rush to collapse differences into continuity. One of the repeated claims in Derrida's late works on animals is precisely to dispute the assumption of autonomy and sovereignty that is traditionally bequeathed to the human over the animal because it ignores the various problems of the drive, force, or play of the unconscious in that human subject, an unknowability which is extended to ideas of the trace or text itself.[23] For this reason, in *The Beast and the Sovereign*, Derrida cautions that a dream of paradise (or what David Farell Krell glosses as 'Peaceable Kingdom") can only ever be a phantasm that elides its own economy of finitude, force, and calculation.[24] The problem of force, and of the play of response and reaction, also helps us to rethink the novel's motif of transition and being "in the approaches" of something. The characters sometimes register this as a mystery of instinct, unconscious thoughts, or even something more transcendental. After a particularly fraught attempt to take a gift to Franklin, Carla speculates on chance:

> Someone more credulous—more given to superstitious thoughts, I suppose—might suspect the involvement of supernatural forces in the seeming sabotage of this uncontentious delivery process. Or perhaps, I pondered (falling back instead, rather more keenly, on the teachings of science), but its merely my unconscious mind trying to trip me up … . But then why …? What could I possibly …? Uh …? (157)

Franklin also rages "TO FEEL THESE … THESE … LUDICROUS THINGS MUST BE THE UNCONSCIOUS EXPRESSION OF SOME KIND OF PROFOUND PSYCHOLOGICAL SELF-LOATHING!" (246). Therefore, the problem of being "in the approaches" to something is also shown to be a problem of understanding a motive or drive or, to rephrase Derrida on the animal again, the problem of following and the response. This describes the experience of the reader as well as the character, and so it is best to think that following animals in *In The Approaches* is not an expression of vitality or energy but more an experience of narrative and trace.

The conclusion of the novel, in which Franklin sees the unveiling of Carla in a vision of light, reveals the competing ideas of life, narrative, and event:

> Because I am exhausted. And I really don't know how this will all turn out. And Miss Hahn is banging away at the window—the window of this story—asking to be let in again. And I shall let her in, soon enough. But first at least let me … let me relish this moment. At the start and the end of something. Let me take my time to decide what kind of carriage—car—receptacle I am inhabiting, and then push the door open—to welcome her. My dear Miss Hahn.
>
> Perhaps I am simply inside my own heart, and I am steering this conversation from within, And the banging of Miss Hahn's fist is the beating of that heart.

> I can't say. All I can really be sure of, right now, is that she's standing there. God bless her. (And so are you—God bless you too.) Pounding and pounding and pounding and pounding. (497)

This final paragraph allows us to consider some competing strains of the novel. The pounding heartbeat that provides a metonymic figure of pounding on a window seems to emphasize the living, throbbing presence of the body, as a sign of energy and vitalism, thus ending the novel of an affirmation of life. However, the same passage could also be read as a condition of suspension, of waiting for the Other, and of an opening to *l'arrivant*. The mysterious "you" that follows Miss Hahn as the addressee of the speech may be a metafictional device to the reader, yet also makes the sentence a performative address to the other close to a prayer (The piety "God bless you" *is* potentially a cloying and bland platitude, but also opens up speech to a relationship to the other, even a form of prayer). As an obligation to the Other, it also resembles the later arguments of *The Beast and Sovereign* seminar, especially the reflection on Paul Celan's line "The World is gone. I must carry You," about how the only possible sense of living in a world depends on such an obligation to bear the other "as if" we were responsible for them.[25] The ending of the novel is less a celebration of life as irrepressible vitalist energy, than this condition of the survival of life as the opening to the future and the promise. This also might account for how the novel explores a "post-secular" condition whereby unconventional faith, institutional religion, and science compete for authority in the modern world.[26] In particular, the novel attempts to conflate Christian belief and symbolism with the invention of mathematical codes and technology as signs of mutual assurance: both faith and knowledge are beliefs that are said to sustain the possibilities of future life. And in the end, it must be said, this carries artistic risks: the novel combines so many topical references, wild coincidences, high speculation, and low comedy, that a reader may find it unwieldy, uneven, or simply odd, or dismiss the spirituality of the fiction as muddled or nebulous.[27] Moreover, for those who want the contemporary novel to expand the representation of animals as beings, the typological mode of the work as presented here may well seem disappointing, reducing the animals themselves into simple signs for human contemplation. It might recall the reservations Derrida makes against Levinas' celebrated paean to Bobby the dog: that the animal that is read as a sympathetic witness to the human is still disavowed from more serious thinking of difference.[28]

There is no getting away from the strangeness of *In the Approaches*, but ultimately it might be more helpful to think of it as an apocalyptic fiction addressing a particular sense of an ending in British culture. As the long historical moment of the Postwar begins to fissure into competing historical periods, and Postmodernism seems increasingly to have become obsolete, this fiction about England in 1984 as an unnoticed transition from one way of life to another seems to be rethinking technology and the digital era as a form of religious fable. For Barker's novel is perhaps ultimately about the possibility of a future as such in the modern fiction of the world. That it uses animals as a way to think of this change is a sign of its singular literary animality, which Derrida's own following of literary animals—even to the very end of the world— helps us to understand.

NOTES

1. Jacques Derrida, *The Animal That Therefore I Am*, ed. Marie-Louise Mallet & trans. David Wills (New York: Fordham University Press, 2008), 12.
2. Tibor Fabiny, *The Lion and the Lamb: Figuralism and Fulfillment in the Bible, Art and Literature* (Basingstoke: Macmillan, 1992), 2. See also, Eric Auerbach, "Figura," in *Scenes From the Drama of European Literature*, trans. Ralph Mannheim (Minneapolis: University of Minnesota Press, 1984), 58.
3. Jacques Derrida, *Spectres of Marx: The State of Debt, The Work of Mourning and the New International*, trans. Peggy Kamuf (London: Routledge, 2006), 82.
4. Derrida, *Animal*, 132.
5. Greg Garrard, *Ecocriticism*, 2nd ed. (London: Routledge, 2012) 153.
6. Frank Kermode, *The Sense of An Ending: Studies in the Theory of Fiction* (Oxford: Oxford University Press, 1967), 58–9.
7. Nicholas Royle, "Jacques Derrida and the Future of the Novel," in *Derrida Now: Current Perspectives on Derrida Studies*, ed. John W. Phillips (Cambridge: Polity, 2016), 193.
8. Peter Childs, *Contemporary Novelists: British Fiction since 1970* (Houndmills: Palgrave, 2–5), 3. See also, Judy Ellam & Tom Wright, "Nicola Barker." British Council: Literature, accessed January 30, 2015, http://literature.britishcouncil.org/writer/nicola-barker.
9. Dominic Head, *The State of the Novel: Britain and Beyond* (Malden MA: Wiley-Blackwell, 2008), 14.
10. Nicola Barker, *In The Approaches* (London: Fourth Estate: 2014), 45–6. Further references to the novel will be given in the text.
11. Sarah E. McFarland, "Animal Studies, Literary Animals, and Yann Martel's *Life of Pi*," in *The Cambridge Companion to Literature and the Environment*, ed. Louise Westling (Cambridge: Cambridge University Press, 2014), 155.
12. Royle, "Future of the Novel," 196.
13. Garrard, *Ecocriticism*, 155.
14. Timothy Clark, *The Cambridge Introduction to Literature and the Environment* (Cambridge: Cambridge University Press, 2011), 179.
15. Jacques Derrida, *The Beast and the Sovereign*, Volume II, trans. Geoffrey Bennington (Chicago: Chicago University Press, 2011), 88.

16. Michael Lucey, "Realism," in *The Cambridge History of French Literature*, ed. William Burgwinkle, Nicholas Hammond, and Emma Wilson (Cambridge: Cambridge University Press, 2014), 466. Flaubert's story is available as *The Simple Heart*, trans. Roger Whitehouse (Harmondsworth: Penguin, 2015).

17. Derrida, *Animal*, 132.

18. Steven Shakespeare, *Derrida and Theology* (London: T & T Clark, 2009), 212.

19. Jacques Derrida, *The Beast and the Sovereign*, Vol. I, trans. Geoffrey Bennington (Chicago: Chicago University Press, 2009), 183. The allusion to *The Triumph of Life* might be to the P. B. Shelley poem discussed by Derrida in the essay "Living On" in *Deconstruction and Criticism*, Harold Bloom *et al.* (London: Continuum, 2004), 62–142.

20. Andrew Stott, *Comedy*, 2nd ed (London: Routledge, 2014), 28.

21. Claire Colebrook, "Not Symbiosis, Not Now: Why Anthropogenic Change is Not Really Human," Th*e Oxford Literary Review* 34, No. 2 (2012): 192–4.

22. Giorgio Agamben, *The Open: Man and Animal*, trans. Kevin Atell (Stanford: Stanford University Press, 2004), 92.

23. Derrida, *Beast*, Vol. I, 130.

24. Derrida, *Beast*, Vol. I, 302. The 'Peaceable Kingdom' is discussed in David Farell Krell, *Derrida and Our Animal Others: Derrida's Final Seminar, 'The Beast and the Sovereign'* (Bloomington: Indiana University Press, 2013), 119.

25. Derrida, *Beast*, Vol. II, 267–8.

26. Irving Goh, "After Derrida's *Foi er savoir*: From Rejection to the (Animal-) Reject for the 'Post-Secular,'" in *Derrida Now: Critical Perspectives on Derrida Studies*, ed. John W. Phillips (Cambridge: Polity, 2016), 132–3.

27. See the review by Lucy Ellman, "In the Approaches by Nicola Barker review-a quirky tale of reconciliation," *Guardian*, June 11, 2014 http//www.theguardian.com. book/2014/jun/11/in-approaches-nocola-barker-review-novel.

28. Derrida, *Animal*, 117.

WORKS CITED

Agamben, Giorgio. *The Open: Man and Animal.* Trans. Kevin Attell. Stanford, CA: Stanford University Press, 2004.

Auerbach, Erich. "Figura." In *Scenes From the Drama of European Literature*, trans. Ralph Mannheim, 11–76. Minneapolis, MN: University of Minnesota Press, 1984.

Barker, Nicola. *In the Approaches*. London: Fourth Estate, 2014.

Brooks, Peter. *Reading for the Plot: Design and Intention in Narrative.* Cambridge, MA: Harvard University Press, 1984.

Childs, Peter. *Contemporary Novelists: British Fiction Since 1970*. Houndmills: Palgrave, 2005.

Clark, Timothy. *The Cambridge Introduction to Literature and the Environment.* Cambridge: Cambridge University Press, 2011.

Colebrook, Clare. "Not Symbiosis, Not Now: Why Anthropogenic Change Is Not Really Human." *The Oxford Literary Review*34, No. 2 (2012): 185–210.

Derrida, Jacques. *Specters of Marx: The State of Debt, The Work of Mourning and the New International*. Trans. Peggy Kamuf. London: Routledge. 2006.

———. *The Animal That Therefore I Am*. Trans. David Wills. New York, NY: Fordham. 2008.

———. *The Beast and the Sovereign*. Vol. 1. Trans. Geoffrey Bennington. Chicago, IL: Chicago University Press, 2009.

———. *The Beast and the Sovereign*. Vol. 2. Trans. Geoffrey Bennington. Chicago, IL: Chicago University Press, 2011.

Ellam, Judy & Tom Wright. "Nicola Barker." British Council: Literature, accessed January 30 2015, http://literature.britishcouncil.org/writer/nicola-barker.

Fabiny, Tibor. *The Lion and the Lamb: Figuralism and Fulfilment in the Bible, Art and Literature*. Basingstoke: Macmillan, 1992.

Farrell Krell, David. *Derrida and Our Animal Others: Derrida's Final Seminar, "The Beast and the Sovereign."* Bloomington, IN: Indiana University Press, 2013.

Flaubert, Gustave. *A Simple Heart*. Trans. Roger Whitehouse. Harmondsworth: Penguin, 2015.

Garrard, Greg. *Ecocriticism*. 2nd ed. London: Routledge, 2012.

Goh, Irving. "After Derrida's *Foi et savoir*. From Rejection to the (Animal-) Reject for the 'Post-Secular." In *Derrida Now: Current Perspectives in Derrida Studies*, edited by John W. Phillips (Cambridge: Polity, 2016), 132–59.

Head, Dominic. *The State of the Novel: Britain and Beyond*. Malden, MA: Wiley-Blackwell, 2008.

Kermode, Frank. *The Sense of an Ending: Studies in the Theory of Fiction*. Oxford: Oxford University Press, 1967.

Lucey, Michael. "Realism." In *The Cambridge History of French Literature*, edited by William Burgwinkle, Nicholas Hammond & Emma Wilson, 461–70. Cambridge: Cambridge University Press, 2011.

McFarland, Sarah E. "Animal Studies, Literary Animals, and Yann Martel's Life of Pi." In *The Cambridge Companion to Literature and the Environment*, edited by Louise Westling, 152–65. Cambridge: Cambridge University Press, 2014.

Royle, Nicholas. "Jacques Derrida and the Future of the Novel." In *Derrida Now: Current Perspectives in Derrida Studies*, edited by John W. Phillips, 184–212. Cambridge: Polity, 2016.

Shakespeare, Steven. *Derrida and Theology*. London: T & T Clark, 2009.

Stott, Andrew. *Comedy*. 2nd Ed. London: Routledge, 2014.

Index

About the Contributors

José Alaniz, Associate Professor in the Department of Slavic Languages and Literatures and the Department of Comparative Literature (adjunct) at the University of Washington, Seattle, authored *Komiks: Comic Art in Russia* in 2010 and *Death, Disability and the Superhero: The Silver Age and Beyond* in 2014 (both published by the University Press of Mississippi). He chairs the Executive Committee of the International Comic Arts Forum (ICAF), the leading comics studies conference in the United States. His current book projects include *Beautiful Monsters: Disability in Alternative Comics* and a history of Czech graphic narrative.

Sarah Bezan is a Newton International Fellow at Sheffield University's Animal Studies Research Centre. Her research focuses on contemporary paleoartistic renderings that engage with the principle of contingency (or chance) that shapes our future on earth in a time of environmental uncertainty. She is also developing a book-length project on the decompositional aesthetics of evolution in the thinking of Charles Darwin, alongside contemporary artists, filmmakers, poets, and authors. Sarah's work has appeared in the *Journal for Critical Animal Studies, the Journal of the African Literature Association, Criterion, and Mosaic: Journal for the Interdisciplinary Study of Literature.*

David Brooks, an award-winning poet, novelist, short fiction writer, and essayist, taught Australian Literature (1991–2013) and directed the graduate writing program (2000–2013) at the University of Sydney where he is now an Honorary Researcher and Associate Professor. Since 1999, he has been coeditor of *Southerly*, the leading journal of Australian literature and new Australian writing. He was the 2015/2016 Australia Council Fellow in Literature. His works include *The Book of Sei* (stories, Faber & Faber, 1987),

The Fern Tattoo (novel, University of Queensland Press, 2008), *The Sons of Clovis: Ern Malley, Adoré Floupette, and a Secret History of Australian Poetry* (literary history/criticism, UQP, 2012), and more recently *Open House* (poetry, UQP, 2015), *Napoleon's Roads* (short fiction, UQP, 2016), and *Derrida's Breakfast* (essays, Brandl & Schlesinger, 2016). A vegan and animal rights activist, he lives in the Blue Mountains of New South Wales.

Megan E. Cannella is a PhD student of literature at the University of Nevada, Reno. Her research primarily focuses on twenty-first century understandings of place and identity, as well as how these understandings have changed the narrative structure of twenty-first century novels.

Her published essays include "The Pain and Prison of Post-9/11 Parenting in Jonathan Franzen's *Freedom*" in the edited collection *Representing 9/11: Trauma, Ideology, and Nationalism in Literature, Film, and Television* and "The Quiet Zoo: The Role of Quiet in Suburban Domesticity in Franzen's *Freedom* and *The Corrections*" in the edited collection *An Introvert in an Extrovert World: Essays on the Quiet Ones*.

Bonnie Gill is an independent scholar and freelance editor and translator living in Charlottesville, Virginia. She received her PhD in French from the University of Virginia in 2016. Her current research focuses on film, trauma, and cultural translation.

David Huebert is a Social Sciences and Humanities Research Canada (SSHRC) Graduate Scholar and PhD candidate at Western University, where he is researching interspecies erotics in American literature. David has published articles in journals such as *Transgender Studies Quarterly*, *English Studies in Canada*, and the *Canadian Review of American Studies*. David is also the author of a poetry collection (Guernica, 2015) and a book of short stories, forthcoming from Biblioasis.

Nicole Mennell is a PhD candidate within the School of English at the University of Sussex, United Kingdom, funded by the Arts and Humanities Research Council (AHRC) as part of the Consortium for the Humanities and the Arts South-East England (CHASE). Her thesis explores the connections between figures of sovereignty and animals in early modern texts, with a particular focus on the ways that animals are used to symbolize the power and status of political leaders, both figuratively and literally, and how this reliance conversely undermines anthropocentrism. Nicole has also assisted on the Wellcome Trust-funded project "'Hidden Persuaders': Brainwashing, Culture, Clinical Knowledge and the Cold War Human Sciences, c. 1950–1990," and recently cofounded an online open access postgraduate journal entitled *Brief Encounters*, which showcases interdisciplinary work in the arts and humanities.

Malin Palani is a Visiting Assistant Professor at Macalester College, where she teaches courses in performance theory, performance art and avant-garde theatre, and performance and ecology. She made her first forays into directing while doing her PhD at the University of Minnesota, taking on shows such as Heiner Müller's *Die Hamletmaschine* and the devised collaboration *The Woyzeck Project*. Her dissertation research examined contemporary performance from the perspective of relationality—theorizing a modality of care and encounter with the more-than-human. She is currently working on a devised production that weaves together her research on the performance of pollutants in waterways and her practice as a visual artist, director, and puppeteer.

Rodolfo Piskorski is a PhD candidate in the Centre for Critical and Cultural Theory at Cardiff University. He has published and presented widely on the interface between textuality and animality.

Gavin Rae is Conex Marie Skłodowska-Curie Experienced Research Fellow at the Universidad Carlos III de Madrid, Spain, where he is the principal investigator for the research project "Sovereignty and Law: Between Ethics and Politics." He was previously an Assistant Professor of Philosophy at the American University in Cairo, Egypt where he also held an Andrew Mellon Postdoctoral Fellowship. He specializes in post-Kantian philosophy with particular emphasis on ontology, socio-political philosophy, and theories of subjectivity. He has published three books, *The Problem of Political Foundations in Carl Schmitt and Emmanuel Levinas* (Palgrave Macmillan, 2016); *Ontology in Heidegger and Deleuze* (Palgrave Macmillan, 2014); and *Realizing Freedom: Hegel, Sartre, and the Alienation of Human Being* (Palgrave Macmillan, 2011), and numerous articles in international journals including *Critical Horizons, Epoché: A Journal for the History of Philosophy, History of the Human Sciences, Human Studies, International Journal of Philosophical Studies, Journal of the British Society for Phenomenology, Journal of International Political Theory,* and *Sartre Studies International*. He has been invited to present his work at international conferences throughout Europe, North America, and North Africa and has also been a Visiting Professor at the University of London (Royal Holloway) and Universidad Complutense de Madrid.

Kirsten Strom is Professor of Art History at Grand Valley State University where she teaches courses on Surrealism and Animals in Art and was a recipient of the University's Pew Teaching Excellence Award in 2012. She is the author of *The Animal Surreal: The Role of Darwin, Animals, and Evolution in Surrealism*, forthcoming from Ashgate/Routledge. She is also an exhibiting and performing artist, composer, and choreographer, who has written on a range of topics including "dance anthropology," serpentine dance films, and postmodern tendencies in the alternative music of the 1980s. Her website is www.barefootballet.net.

James Tink is Associate Professor in the Department of English Literature, Tohoku University, Sendai, Japan. He researches and publishes on early-modern literature and drama, contemporary British literature, and literary theory. He is currently working on a project about temporality and ideas of labor in English literature.

Lightning Source UK Ltd.
Milton Keynes UK
UKOW06n0835231117
313207UK00015BA/703/P